OTHER DADANT PUBLICATIONS

BEEKEEPING
Questions and Answers

Edited by
DADANT & SONS

FIRST EDITION

DADANT & SONS • HAMILTON, ILLINOIS

Second Printing 1979

Library of Congress Catalog Card Number 77-80061

ISBN Number 0-915698-04-8

PRINTED IN THE U.S.A. BY JOURNAL PRINTING COMPANY

CARTHAGE, ILLINOIS

PREFACE

Honey bees and beekeeping are steeped in ageless time. Numerous cave paintings depict primitive men and women robbing honey from the nests of wild bees in the crevices of cliffs. The desire for honey led to the keeping of bees and beekeeping was already a mature art by the time it was heralded in the Koran of Mohammed and the Holy Bible of Christianity. Beekeeping was mentioned in Greek mythology, it was studied by Pliny and Aristotle, and the honor role of researchers and inventors has continued to this day, illuminated here and there by such names as Dzierzon, Huber, Fabre, Quinby, Langstroth — the list is endless.

On January 1, 1861, the first issue of *The American Bee Journal* appeared, the first bee magazine to be published in the English language. As Frank C. Pellett was to say in his *History of American Beekeeping,* "The history of *The American Bee Journal* has been the history of the rise of beekeeping, and the one is inseparably linked to that of the other."

Looking back on the status of beekeeping on that January day in 1861, the industry was as remarkable for what it did not contain as it was for the progress that had been made. Communication was slow and troublesome and beekeepers had not begun to hold meetings and conventions. Men labored over their research and inventions, sometimes only miles apart, without knowing of the work and progress of the other. A practical bee smoker had not yet been invented, queen excluders were still unknown, and comb foundation was still to be perfected. The extractor had not yet come into use and commercial queen rearing had not been suggested.

The Rev. L. L. Langstroth had discovered the principle of the "bee space" and had invented his hive, two achievements which were to rightfully earn him the title of "The father of modern beekeeping," but his hive was not yet in common use. Most of the important inventions, as well as discoveries in management techniques as now practiced by the beekeeping industry, occurred after the birth of *The American Bee Journal* and were first brought to the attention of the public in its pages.

When Samuel Wagner, the Journal's founder, announced his plans for the magazine, he said, "It will serve as a repository of whatever is of practical value in this department of rural economy; and as a vehicle by which information can be readily, rapidly and widely diffused so that the early introduction of useful improvements may be secured."

The American Bee Journal fulfilled that promise and future pages were to record the invention of the extractor, the smoker, comb foundation, the introduction of alsike clover and sweet clover, and the movement to bring about enactment of a pure food law. The Italian bee was brought to the attention of beekeepers through its pages and several men whose names were to become household words in the industry, began their careers in the Journal. A. I. Root began his career as a contributor to the Journal, writing under "Experiences of a Novice in Beekeeping" for many years. Charles Dadant made his first contribution in November of 1867, introducing himself as a newcomer from France.

In October of 1870, Dr. C. C. Miller first appeared as a contributor and from that time on until his death in 1920, served as associate editor and helped establish the question and answer section as one of the most popular features of the magazine.

When Samuel Wagner died in 1872, the Journal continued under various editors including Thomas G. Newman and George W. York until 1912 when it came to Hamilton, Illinois, under the editorial direction of C. P. Dadant. C. P. Dadant died in 1938 and *The American Bee Journal* continued under the editorship of G. H. Cale, Frank C. Pellett, and M. G. Dadant, joined by Roy Grout, another member of the Dadant family, in 1945.

This book is very properly a child of *The American Bee Journal.* It was written over a period of several years by a variety of people under everchanging circumstances. The text is composed of questions submitted to *The American Bee Journal* and the answers we gave. Some of the questions are from our back files and appear in print for the first time and some of the answers have undergone changes as beekeeping knowledge, equipment, and management techniques have also undergone changes. Duplicate questions have been avoided wherever possible.

The questions and answers range from the simple to the complex and they came from beekeepers, large and small, both experienced and beginner, from nearly every known spot in the world where bees and beekeepers live. The only common denominator was that they all contained problems and we tried to help.

It has been estimated that there has been more written about bees and beekeeping than any other topic with the exception of religion and yet, there are still perhaps many more questions than answers.

These particular questions and answers have been compiled and edited into book form in the hopes that they may answer your questions about bees and beekeeping.

Howard Veatch
Dadant & Sons, Inc.

TABLE OF CONTENTS

BEES

QUEEN MOST IMPORTANT IN BEEKEEPING

Q. What do you consider the most important thing in all beekeeping, if you can consider one thing more important than others?

A. A thorough knowledge of everything connected with the business. Perhaps you want to know which is the most important, the bees, pasturage, hive, or some other thing. Hard to say, bees are no good without pasturage, and pasturage is no good without bees. You can't very well get along without a hive. But if we were to pick out some one thing to which the beekeeper must give the greatest attention, we would say the queen. For whatever the queen is, that decides what the bees are. By breeding for the best all the time, a man is more likely to get ahead than by giving his attention to something else, such as hives or pasturage.

Fig. 1. This vigorous young hybrid queen is the product of careful selection for gentleness, egg-laying ability, and her off-spring will also be excellent honey-gatherers. (Photo, E. R. Jaycox)

THE HONEY BEE AS A COLD-BLOODED INSECT

Q. The Book of Knowledge states that a honey bee is a cold-blooded insect. I always thought that if they maintained a temperature above that of the surrounding air they would be rated other than cold-blooded.

A. We would be of the opinion that the Book of Knowledge is correct when it states that the honey bee is a cold-blooded insect. Webster's dictionary defines cold-blooded as a creature having a variable body temperature, not internally regulated.

The honey bee colony is able to regulate the temperature of the hive at 93°F. in summer and 57°F. in winter by consuming food and clustering together to keep the hive warm and by ventilation with their wings in summer. They also evaporate moisture within the hive to drop the temperature. However, a single honey bee would die on a cool night because its body temperature would drop to that of the surrounding air quickly. One might say then that the honey bee colony is warm-blooded while the individual honey bee is a cold-blooded insect.

LIFE-SPAN OF QUEEN, DRONE, AND WORKER

Q. What is the average life-span of a queen, drone, and worker?

A. This answer is going to have a million possibilities for being wrong, considering conditions in and around your hive, accidents, and the other variables that affect bees and beekeeping, so these figures represent average life-spans.

A queen has a high risk factor considering the rivalry for her position at the head of the colony, but depending upon conditions and barring accidents, she might live two or three years. She could probably survive up to four or five years, but her life is generally ended at the same time her productivity fades, so two or three years is fairly safe. A wise beekeeper will forget averages and keep a close eye on the productivity of the queen in each colony, requeening as the need arises to keep the colony in peak condition for the working season.

Worker bees literally wear themselves out with the work of the colony and six weeks or perhaps less in the working season is about average. Workers live through the relatively inactive winter months and die intermittently during the early spring brood rearing period when young bees are emerging to replace them.

Drones lead a much more hazardous life and one working season or less is all they can count on. If a drone is successful in mating with a virgin queen, his life is ended and if the drone does live on through the season, the workers will drive him out of the hive in the fall to meet a certain death from cold and starvation.

SENSE OF COLOR

Q. How does a bee's sense of color compare to ours?

A. Dr. Karl von Frisch has conducted many experiments to find out whether bees have a color sense and his results have been published in *Bees, Their Vision, Chemical Senses, and Language,* available through most beekeeping supply houses. Dr. von Frisch found that bees do have a true color sense as he was able to train them to come to food on blue, orange, yellow, green, violet and purple cardboard. However, when he tried to train bees to come to food on red cardboard, they also landed on the black and dark gray cards, indicating that bees are red-blind. In other tests where bees sometimes confused yellow with orange and green and blue with violet, it was concluded that bees apparently see yellow, blue, green, and ultraviolet as four different qualities of color.

To compare the color sense of bees and man, the visible spectrum is shortened for bees in the red but it is extended into the ultraviolet. Man can distinguish red which the bees cannot and bees are able to distinguish ultraviolet colors that we cannot see. While bees distinguish only four different qualities of color, the average human can distinguish about 60 distinct colors in the visible spectrum and a trained human eye such as that of a decorator might be able to distinguish perhaps 70 or more colors. The comparison of bees to man, as in most cases, is rather futile since bees have a more narrowed interest in color (related only to food collection) while man surrounds himself with color for aesthetic reasons. Obtain a copy of Dr. von Frisch's book and read it as it is a pleasure to read and very helpful in attempting to understand the activities of honey bees.

SENSE OF TASTE

Q. Do bees really have a sense of taste and if so, why would they collect materials for which they have no real use?

A. Dr. von Frisch has done the most experimental work in this field and my answer to your question is based upon his findings. According to Dr. von Frisch, bees are able to distinguish sweet tastes, as we might suspect from their affinity with honey, but can also single out salty, sour, and bitter tastes as well. Honey bees were found to be slightly more sensitive to salty and sour tastes than humans but less sensitive to bitter tastes than humans as they readily took a mixture of quinine and sugar which would be totally disgusting to the human sense of taste. Bees were also found to be sensitive to sugar solutions as low as one or two per cent which is considerably more acute than the human sense of taste.

In answer to your question of why bees sometimes gather materials such as sawdust, coal dust, and various animal feeds for which they likely have no real use, Dr. von Frisch also found that hungry or starving bees lowered their thresholds of acceptance in times of need much as a hungry child eats his vegetables better than one with no appetite. Your bees are probably in need of more pollen than local floral sources are supplying. We suggest that you feed a pollen supplement as hunger is driving your bees to collect these materials they will not be able to use.

COMMUNICATION AMONG BEES

Q. In beginning my study of bees, I have come to understand that they have a rather complicated communication system and that one returning forager can tell the rest of the bees in the hive where the nectar and pollen sources are located. Can you tell me how this works?

A. Quite a bit of work has been done in the area of bee communications and there is much that we still do not understand. We're going to summarize some of the work that has been done but we're also going to refer you to two books for a complete survey of the work in this fascinating field: *The Hive and the Honey Bee,* a Dadant & Sons, Inc. publication and *The Dance Language and Orientation of Bees* by Dr. Karl von Frisch.

Dr. Norman Gary reminds us in his chapter in *The Hive and the Honey Bee* that bees are social animals and that effective communication is a basic requirement of a social existence. He also points out that we tend to interpret bee activities and behavior by ascribing to them human characteristics or terminology. We cannot assume that there is a "thinking process" or "intelligence" in bee communication as we understand these terms.

Bees use various stimuli, such as light, chemical, and physical stimuli, that can be perceived by specific sensory organs just as communication signals are sent and received by human beings. The honey bee is adapted to the relatively close and dark atmosphere of the interior of the hive suggesting that

Fig. 2. Food-sharing is an important means of communication for bees and they often "beg" for honey even though standing on open cells of honey from which they could easily help themselves. (Photo, Bee-Wolf)

odor, touch, and perhaps sound, rather than vision, may be the primary stimuli involved in their communication.

A foraging bee may return to the hive and pass her drop of nectar to receiving bees in the hive that are potential foragers. This sharing of food is probably significant in alerting the hive bees that there is food to be collected. There is also evidence that the odor from the fragrance of the visited flower clings to the waxy cuticle layer covering the body of the bee and this odor further serves to identify possible food sources.

Nectar and pollen collectors, upon returning to the hive, may "perform" on the combs in patterns which have been termed "round dances" or "wag-tail dances." These movements excite other bees which follow the dancing of the "performer" with their antennae positioned on or near the dancing bee. The dancing bee may suddenly stop the dance and move to a different place in the hive and resume dancing there, thus exciting other bees. These excited bees are then stimulated to leave the hive in search of the food source. In other dances, especially the wag-tail dance, the "performer" is also able to indicate the direction of this food source in relation to the sun. By a series of movements, the exact angle toward or away from the sun can also be communicated. The quantity of food available is probably indicated by the number of "dances" performed in the hive. A short foraging trip would indicate a heavier honey-flow and there would consequently be more dancing in the hive and more stimulation to fly to the food source.

There are probably many other communication signals that we are not aware of at this time or we may be misinterpreting communication signals that we observe. It has been suggested that electrostatic charges may build up on the body of the foraging bee and these charges may be passed on or detected by other bees, stimulating them to forage.

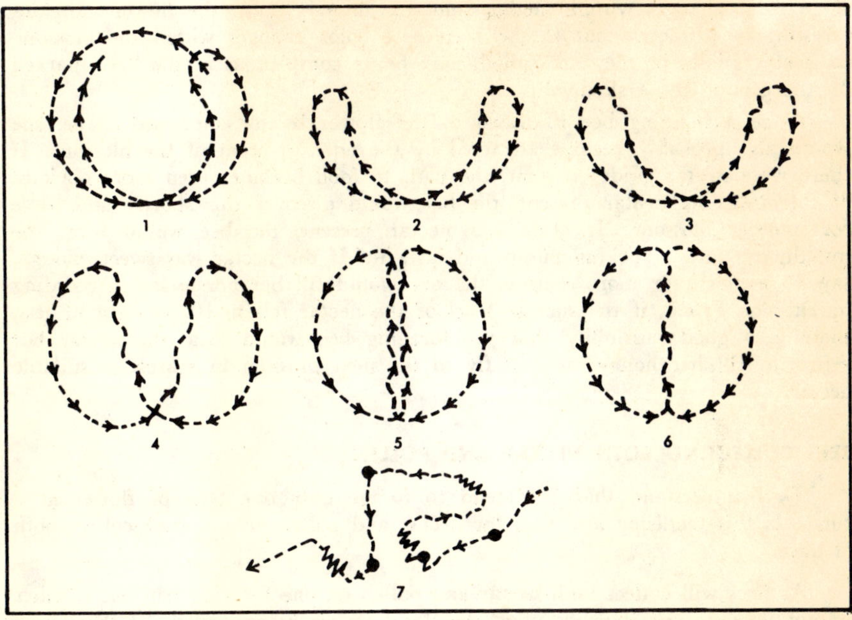

Fig. 3. Honey bee dances (diagrammatic). Round dance (1). Sickle or crescent dance (2). Transitions between the round dance and the wag-tail dances (3-5). Wag-tail dance (6). Pull dance (7).

BEES FINDING NECTAR

Q. I am intrigued by the idea of bee "dances" and wonder, does a foraging bee only follow the information received in the hive or does she have other means of finding nectar sources?

A. Although we are never sure exactly what motivates a foraging bee, it would be safe to assume that the bee goes beyond what information may have been received in the hive.

Dr. Karl von Frisch discovered in his experiments that bees are probably guided by a combination of things: color of the flower, the shape and scent of the blossom, nectar guides within the flower, and the sweetness of the nectar itself. Let us assume that a foraging bee may have received information in the hive as to the location of a floral source. Although it has been discovered through experiments that the information relayed to the hive is remarkably complete, it is doubtful that the bee knows exactly how many feet and inches to fly to reach a given flower.

Dr. von Frisch discovered that the foliage, which is green and attractive to us, is an almost colorless gray with a weak yellowish tinge for bees. The colorful flowers must stand out very effectively against this neutral background. Further experiments showed that the flowers depending upon bees for pollination and those which yielded nectar were rather conspicuous in the qualities of color that bees are known to see. Many of these flowers also contain a contrasting color on or near the entrance to the tube containing the nectar. To us, these color changes within one blossom may merely make the flower more interesting or attractive but to the bee, these color changes within the blossom, or nectar guides as they are called, may be as conspicuous as the "X" marked on the ground for a skydiver.

Once a foraging bee is drawn to the flower by its color and shape, she would also probably become attracted by the odor or scent of the blossom. If there were nectar guides present, the path to food becomes even more obvious. If there was no nectar present, the bee would discover this shortly and leave for another blossom. If there was nectar present, the bee would form the mouthparts into a sucking pump and taste it. If the nectar was sweet enough, say 20 per cent or more sucrose, the bee would fill her honey sac. According to Dr. von Frisch, if the sucrose level of the nectar fell to 10 per cent or less, there is a good possibility that the foraging bee would taste the nectar but refuse to fill her honey sac and fly to the next blossom in search of suitable nectar.

BEES COLLECTING BOTH NECTAR AND POLLEN

Q. I understand that bees tend to forage only one type of flower at a time. Is this tendency also true for nectar and pollen or will they collect both at once.

A. Bees will collect both nectar and pollen on one foraging trip but whether or not they do this depends upon the floral source being visited. If the flower which is being visited is secreting nectar as well as providing pollen, a considerable number of the foragers took both in studies conducted by R. L. Parker. He found that of more than 13,000 bees being observed, 25% were gathering pollen, 58% nectar only, while 17% were collecting both nectar and pollen on the same trip.

PHEROMONES

Q. I have recently read of pheromones affecting bee behavior. How do they affect bee behavior and are pheromones the same thing as what we used to call "queen substance"?

A. Pheromones have been defined by Dr. Norman Gary in *The Hive and the Honey Bee* as "chemical messengers" that are secreted externally and elicit behavioral or physiological responses. Queen substance, also known as 9-oxo-2-decenoic acid, is one of the best known of these pheromones and is produced in the mandibular glands of the queen bee. This powerful compound becomes spread all over a queen's body, probably when she grooms herself and is groomed by her court of attendant worker bees. This pheromone is transmitted then from worker to worker, probably through food-sharing, and, provided the worker bees receive adequate amounts of this pheromone, they are inhibited from a variety of activities such as construction of queen cells and other queen-rearing activities. This pheromone in adequate amounts also inhibits the growth of worker ovaries and serves as a chemical signal that attracts drones to virgin queens on their mating flights. Still other queen pheromones attract workers to the queen and stimulate feeding of the queen. When a colony fails to receive adequate amounts of these pheromones, as in the loss of the queen, the workers soon begin to rear a new queen, often within 5 or 6 hours.

It has also been found that a cotton pad, saturated with this pheromone, inhibited queen-rearing activities in a queenless colony.

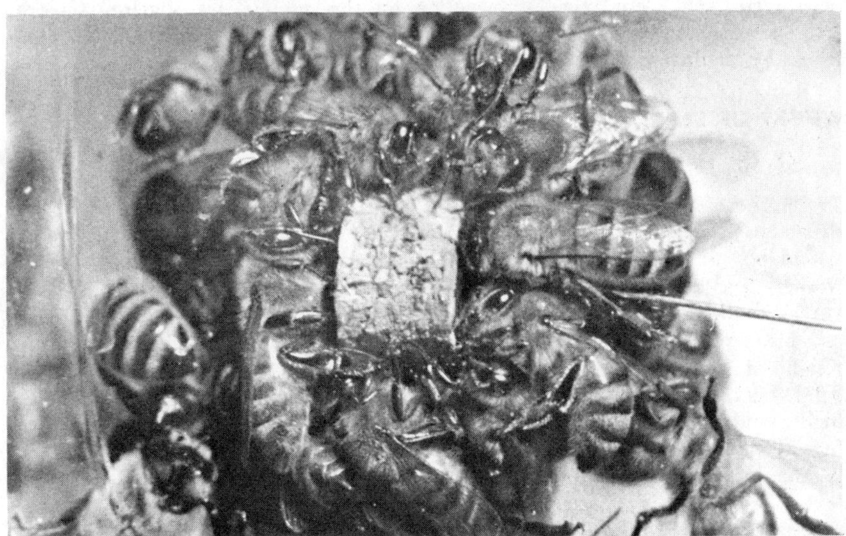

Fig. 4. A cork which has been scented with queen substance is surrounded by workers, drawn to the powerful pheromone.

QUEEN SUBSTANCE

Q. In a recent statement about honey bee communication you said that if the queen was removed and queen substance was introduced into the hive, no queen cells would be formed because the workers would think that the queen was still there. I disagree with you. What are you going to do with her court? You're not going to remove the queen without them knowing it. What queen's substances are you going to put in the hive and where?

A. The queen substance is introduced into the hive by putting some of the material on a cotton ball or blotter and placing it at some point in the hive. The bees do actually form a queen court around the queen substance acting as if the cotton ball or blotter was their queen. This has been demonstrated a number of times under laboratory conditions.

The difficulty here may be that we are accustomed to thinking of the bees from the standpoint of our own intelligence and ability to think through situations. If that were the case, it would make sense that the circle of attendant bees would miss the queen and would not be fooled by the queen substance on the cotton ball, and might even pass the word on to other worker bees. We have not been able to demonstrate that honey bees have an "intelligence" as we know it, or an ability to think as we understand the thinking process. As we understand it, you might say that bees are engineered by nature to react in certain ways to certain stimuli and the queen substance is one of the most powerful stimuli affecting bees.

According to Dr. Norman Gary, Professor of Entomology at the University of California at Davis, "The successful manipulation of bee activities by using synthetic pheromones is an exciting possibility that awaits future research and development. Pain (1973) has listed 31 pheromones, of which 13 have been identified." For more information on this subject, we suggest you read Dr. Gary's chapter on "Bee Behavior" in *The Hive and the Honey Bee,* published by Dadant & Sons, Inc.

WEIGHT OF BEES

Q. On an average, how many bees will it take to make a pound?

A. Let's assume you mean worker bees. Queens and drones are larger than the workers and it would take correspondingly fewer bees to constitute a pound but as you will be buying a 2-pound package that will contain mostly workers and a few drones, let's work from that basis. The weight will also vary with the race as the size of different races varies.

Mr. W. J. Nolan conducted studies in which he found that the average weight of samples taken from 45 packages was 135.85 milligrams per bee or 3,337 bees per pound. Other studies have determined average weights both higher and lower than that but as your package bees were probably pretty well fed at the time of packaging or just after, an average number of bees per pound of package bees is probably somewhere in the neighborhood of 3500.

FLIGHT OF A BEE

Q. Is there any difference in the flight of a loaded bee and one traveling empty?

A. According to the late Dr. O. W. Park of Iowa State University, the average speed of flight of a loaded bee is 15 miles per hour, while the average speed of flight of an empty bee averages 12.5 miles per hour. This suggests the possibility that a bee on her outward journey may not in all cases make a so-called "bee line" for the source of supply, but may sometimes do scouting on the way.

The only case in which the outgoing bee made better time than the incoming bee was when flying directly against the wind. It appeared that, when going with the wind, the bee showed a tendency to slacken her own efforts; whereas, when traveling against the wind, she increased them in an attempt to overcome the wind's retarding influence.

BEES AND FRUIT

Q. Close neighbors claim the bees are eating their grapes and peaches, and one man said they destroyed all of his peach crop. The birds are plentiful here and I tried to explain that it is the birds, but they won't listen to me and want me to move the bees. Can you help me?

A. Yours is not a localized problem. The Dadant family here in Hamilton went through this same problem themselves many years ago. Finally, to prove their point, they planted some large acreages of grapes just to show the skeptics. It was expensive but the experiment proved that bees will not, as some people have thought, sting or bite grapes or any other fruit and draw out all of the juices. The question has also been asked, "If bees can gnaw through a newspaper, why are they unable to cope with the delicate skin of a grape?"

Newspaper is made up of small units or fibers and is rough under the microscope, so the bees are able to catch hold of these little particles with their mandibles and tear them away. They are also highly motivated to do so, usually to escape from one container into another or to release a queen whose pheromones are quite attractive to them. Bees cannot catch hold of absolutely slick surfaces under any circumstances and the skins of grapes and most fruits are far too slick for bees to handle.

Sometimes, after birds or other insects have broken the skins of fruits and if there is little or no nectar available from their natural sources, bees may work on the fruit juice oozing from the damaged fruit.

CARRYING OUT LARVAE

Q. What causes the bees to carry larvae out in front of the hive?

A. There are several causes. A very common cause is the colony's running short of stores and the brood consequently dying of starvation. When wax moth larvae tunnel through the center of a brood comb, they will puncture the bee larvae cells, causing the larvae to die. It then will be carried out by the bees. The same results are brought about by brood that has been either chilled or overheated. When we do have a warm spell in the spring, often the queen will expand her brood nest rather rapidly; then a cold spell comes, the bees cannot cover the brood and it will be chilled. Brood frames that are taken from the hive by inexperienced beekeepers and exposed to the chilled winds for a length of time can cause the brood to be chilled to the extent it will die and be carried out. Another reason for this is a condition resulting from bees gathering a poison substance, feeding it to the larvae and causing the larvae to die. This does not occur often since the bee gathering the poison usually will not be able to get back to the hive alive.

COMMOTION IN FRONT OF THE HIVE

Q. On warm days in early spring I notice much commotion in front of the hive. What is going on?

A. On a bright day after inclement weather, bees which have been confined for some time will often cause an unusual commotion in front of the hive. Bees will be flying in all directions. Watch them for a short time. If concentration of bees remains in the area of the hive, it is probably a play flight. If the colony should be swarming, the concentration of bees in the air moves away from the hive area. If the bees are grabbing each other by the wings and feet and trying to sting each other, it could be robbing. The concentration of bees will be heaviest at the entrance to the hive. If it is robbing, then reduce the entrance to the hive to 2 inches by ½ inch high. This will enable the guard bees to protect their hive better. Do not smoke the hive when robbing is in progress as the smoke will drive the guard bees away from the entrance. If you reduce the entrance and bees are still trying to rob, place some loose grass around the entrance to assist the guard bees in protecting their hive.

Fig. 5. A drone is escorted out of the hive in the fall. The winter stores of the colony will be needed to ensure survival through the coming winter. (Photo, Bee-Wolf)

CARRYING BROOD AND YOUNG BEES OUT OF THE HIVE

Q. Will you tell me what is the matter with my hive of bees this fall? They seem to be working O.K., but they bring out brood or young bees not hatched out or fully developed. I also find some fully developed bees in front of the hive in the grass. They cannot fly or get up out of the grass. The wings seem too small.

A. It is normal for a colony to carry out brood or immature bees in the fall of the year when there is no honeyflow. The adult bees are probably drones which are eliminated every fall in a queenright colony. The bees do not want to feed either the drones which are worthless to a good colony, or drone larvae which will not be needed. It is the instinct of the honey bee to build up population when they will be needed to gather nectar and pollen and then reduce the population again when the surplus has been gathered and stored to enable the colony to form a compact winter cluster and survive until the next season.

ORIENTATION FLIGHTS

Q. When a hive of bees in the afternoon (3:00 to 4:00 p.m.) starts playing around the hive like a small swarm 4 feet wide and 6 feet high, what does this mean?

A. The play flights that you were observing between 3 and 4 in the afternoon were undoubtedly orientation flights by the young bees. This gives the young bees who are just learning the field work a chance to become oriented to the location of the hive so they can find their way back later.

RACES OF BEES

Q. What are the names of the races of honey bees?

A. There are four races of well-known economic importance, all belonging to the family *Apis mellifera* L. Dr. F. Ruttner, in his chapter in "The Hive and the Honey Bee," describes them as follows: Dark Bees (*Apis mellifera mellifera* L.) which are found in all of Europe north and west of the Alps and central Russia; Italian Bees (*Apis mellifera ligustica* Spin.) which originally came from Italy to the United States soon after 1859 and is the race commonly used in this country; Carniolan Bees (*Apis mellifera carnica* Pollmann) whose original homeland was the southern part of the Austrian Alps and Yugoslavia; and Caucasian Bees (*Apis mellifera caucasica* Gorb.) whose original homeland was the high valleys of the central Caucasus. There are, of course, certain minor races and transition forms or strains but these are the principal races of honey bees.

Fig. 6. a b c
Three of the most common European Races of Bees. a. Dark Bee (Apis mellifera mellifera L.) b. Italian Bee (Apis mellifera ligustica) c. Carniolan Bee (Apis mellifera carnica).

RACES OF BEES

Q. I notice in the bee magazines that in addition to the Italian and Caucasian bees which have been the stand-bys for years, the breeders are now offering Carniolan and a number of different hybrids including Dadant's Starline Hybrids.

I would appreciate information as to the "races" of bees now being offered in the advertisements. I am contemplating just a small-time operation to begin with, and I will be teaching at least my own boys — perhaps a group of boy scouts. If there is a race of bees that have the desirable characteristics of the Italians in resistance to foulbrood and the other desirable qualities of this race plus the gentleness of the Caucasian bees, it would seem that these might be almost ideal for my purpose.

A. There is almost as much difference between the different races of bees as there is between the races of anything else. The Italian is by far the most popular race of bees. They generally have a good reputation for gentleness, heavy honey production, non-swarming tendency and other desirable traits. The Caucasian is, as a rule, a gentle race, hardy and productive. It is more difficult to obtain queens since there are fewer breeders and little research has been done on this race as compared to the Italian. Their main problem seems to be the use of excess propolis.

We would strongly recommend the hybrids such as Dadant's Starline, Midnite, or the new Cale hybrid. These hybrids are backed by years of careful research to bring out the best of several lines of bees.

VARIETIES OF ITALIAN BEES

Q. I see advertised at least four varieties of Italian bees: the yellow Italian, the dark Italian, the three-banded Italian, and the leather-colored Italian. What is the difference in relation to production, gentleness, and disease resistant qualities?

A. The question you raise is a good one and one on which many people have very definite ideas. In an organism like the honey bee where natural mating operates as it does, these color variations occur. The breeder can pretty well control the color he wants in his bees, but unless he pays strict attention to other bee characteristics, the variation of a characteristic such as gentleness could be great. In other words, bees selected primarily for their color could exhibit wide variations in other characteristics such as honey production, temper etc. So there is no good reason why the color of the Italian variety should be associated with other general bee characteristics. This is not to say that a particular breeder might not select for other specific characteristics in addition to color. It just means that you cannot associate color varieties with other desirable characteristics and be sure of being right.

HONEY YIELDS FROM HYBRID BEES

Q. I have read of a 34% increase in honey yields of hybrid bees over the average of inbred strains. Is this the maximum, minimum, or average increase? Also, how do bees from queens "bred from the best," which don't have the egg hatchability problem of inbred bees, compare with hybrids?

A. The 34% increase in honey yields cited was actual data from tests made over several years and using several check stocks. Actually, the check stocks were not the inbred ones from which the hybrids were made, but rather were various common stocks commercially available at the time.

In a breed from the best situation, quality of queens will vary from season to season. So will weather conditions which influence the quality of the queen herself. Therefore, any test of this sort would require several years to be valid.

Over the years, the figure of around 30% has come through as the yield advantage for a properly reared and mated hybrid. However, locality and beekeeper management methods do influence this, and no one is sure to what extent. Certainly, a hybrid bred for fast spring buildup would not be able to express itself if buildup space in the brood nest was not provided. Therefore, and to solve the question in your own mind in your locality and under your management methods, we always like to recommend that the beekeeper do his own comparative tests over at least two years. The best way to do it is to order your normal common stock source queens for half of the colonies in an apiary, and the hybrid stock for the other half. Hopefully, more than two yards will be used, and as I mentioned, over more than one season. This, then would give you a good idea of how they would perform in your management system.

QUEENS UNABLE TO LAY

Q. Do some queens become so chilled or injured that they don't even lay drone eggs?

A. Not only can queens be injured by chilling, but physical injury can also prevent them from egg laying. Occasionally you will find a malformed queen that cannot mate or lay eggs.

QUEENLESS HIVES

Q. Is it true that queenless hives make more of a buzzing noise than a queenright hive?

A. A queenless hive will make a loud buzzing noise where a queenright colony does not. The sharp buzzing sound is detectable as soon as you open the cover of the hive. Once you have heard it, you will recognize it quite readily.

STARLINE - MIDNITE HYBRIDS

Q. I am a beginning beekeeper, having started only last spring. I am thinking of requeening with hybrid queens but am confused by the ads for the Midnite and Starline hybrid bees. What is the difference? Why would one be preferred over another?

A. The Midnite hybrid is a scientific hybrid combining four inbred lines of Caucasian origin. The bee has a black background with grey hairs giving a grey-black appearance and it is an extremely gentle bee. It was designed for the hobby beekeeper, but its honey production is good enough that many commercial beekeepers are now making the switch. The Midnite winters well, and has a slightly slower spring buildup than the Starline hybrid. This seems to make it better suited for an area where the honeyflow is not intense, but longer in duration.

The Starline hybrid is based on a four-way cross from inbred lines derived from the Italian race. Although a very gentle bee, it is not as docile as the Midnite. The bees build up rather quickly in the spring, and it seems to do best where the honeyflow is of intense proportions and the buildup is timed perfectly to match with it. The Italians are also inclined to collect little propolis.

STINGLESS BEES

Q. Can you tell me where I can get stingless bees? What is their name?

A. The stingless bee, also known as Meliponidae, always seems to attract some attention because of the "stingless" tag. This doesn't mean that Meliponidae is defenseless. Listen to this variety of weapons: smearing the victim with honey regurgitated from their honey stomachs, pulling hair, running in and out of the eyes, ears, nose, over the hands, and down the shirt collar, and as if this were not enough, biting. They are also without commercial value due to their complicated method of storing honey.

"RUNNING" BEES

Q. I ran into a problem for which I do not find the solution in the books. I decided to transfer a frame of brood from one of my strong hives to one that has not built up fast enough to please me. When I opened the strong hive the bees came spilling out all over the hive. Soon the front and side of the hive were covered with bees. They were not angry as they did not offer to sting. They just wanted to crawl out. I have never had this happen before and just wondered.

A. Bees will "run" for many reasons: too much smoke, starvation, careless handling, but usually it's just the nature of certain strains of bees. Your problem can be solved by requeening, if you can locate the queen in the colony. Some bees are just naturally nervous and will run at the slightest disturbance. They may produce a very good crop in spite of this tendency but it is going to make the hive examinations more difficult and requeening would pay.

WATER-GATHERING ACTIVITIES OF BEES

Q. During a recent hot spell the bees were very seldom around their usual watering spot? Others have noted this at their locations too and they were curious as I am. I would think that the bees would be using more water than usual during the hot weather.

A. We have an answer for you but would also like to point out something in your question that gets in the way of our understanding of bee behavior. We use more water during hot weather and logically assume that bees must need more water also. However, bees use water much differently than we do. Bees use water only in relation to the humidity within the honey bee colony. If they happen to have a lot of nectar in the colony that has a high moisture concentration, then they utilize much of the moisture from this nectar and do not need to fly for water at all even in very hot temperatures. On the other hand, the temperature might be extremely high and yet the humidity might be high at the same time. Under these high humidity conditions, the colony is quite unlikely to need water in spite of the fact that it is hot. In fact, the reverse might be true and the colony would instead need to be trying to dry some of the moisture out of the hive. We believe this would help explain why your bees did not necessarily fly for water at their normal rate even during the hot spell of weather. Honeybees operate so differently from humans that we are often wrong when we attempt to explain or guess their behavior patterns in relation to our own behavior.

REMOVING BEES FROM COMBS

Q. What is the best way of removing bees from brood combs or combs in the honey supers?

A. You could either shake the bees off, use smoke, a bee brush, chemicals, a bee escape, or a bee blower. The method would depend upon your choice, how many combs must be handled, and other factors of your beekeeping operation.

If only a few combs are to be handled, shake the bees off. You probably have picked up the frame with your fingers under the projecting ends of the top-bars, leaving the palms of your hands above the top-bar. Holding the frame in this manner, lower the frame rapidly and jerk it back up to the same level and you will probably shake off most of the bees. The remainder of the bees clinging to the combs may be easily brushed away without injury with a bee brush.

When removing bees from supers, use your smoker or place the bee escape in the center hole of the inner cover, thus converting it to an escape board. Place this escape board under the supers to be removed and recover the hive, making sure that there are no cracks in which robbing could get started as the bees will not be able to return to the supers to guard the honey. By morning most of the supers above the escape board should be free of bees. The few bees that may have remained behind with the honey can be brushed off the combs with the bee brush. A bee blower with its constant stream of air is also of tremendous help in clearing supers of bees if your beekeeping operation is large enough to justify the purchase of a bee blower. Chemicals have generally declined in use although beekeeping supply houses still carry chemicals used to drive bees from supers.

BEESWAX

ORIGIN OF BEESWAX

Q. What is beeswax, or what does it originate from?

A. Beeswax is a substance secreted from eight wax glands located on the underneath or ventral part of a bee's abdomen. It appears as tiny translucent white scales or oval-shaped plates. The wax is produced from the food eaten by the bees, somewhat as a cow produces milk. During good honeyflow periods, when the bees are building new combs, the wax scales are easily seen on the underside of the young, producing bees.

HONEY INTO BEESWAX

Q. How many pounds of honey does it take to make one pound of beeswax?

A. For a long time it was counted 20 pounds of honey to make one pound of beeswax. Whitcomb determined that 3.8 kg. or 8.36 pounds of honey would result in one pound of beeswax. Probably 8 to 10 pounds of honey to make one pound of beeswax would be a safe average.

SURPLUS BEESWAX

Q. Since it would seem no established beekeeper produces enough beeswax to have worked into his necessary foundation, where does the surplus come from?

A. You ask a good question and there are many times when it seems as though every scale of beeswax is in use and there is no surplus. An established beekeeper may not produce enough wax for his own foundation, and again he may. If he works for extracted honey and has reached the point where he makes no more increase and needs no more combs, he may have a surplus of wax from his cappings. Even if he renews his combs, the melted combs should furnish wax for the new ones. It is this sort of situation that supplies the beeswax for the comb honey producer. Surplus beeswax is also produced from melting damaged or moldy combs.

In some of the African countries as well as some other nations of the world, surplus beeswax is produced by salvaging the natural combs that were destroyed when the whole open-air colonies in trees or other nesting sites were raided for honey.

HANDLING BEESWAX

Q. With the season advancing rapidly and the honeyflow in most of the country about to begin, I would appreciate some pointers about handling beeswax.

A. Beeswax is not produced in such large quantities as honey, but because of the higher market price it demands, it does deserve special attention. This refers not only to the methods used in melting down cappings and comb and preparing the cleaned cakes for market, but also to efforts at salvaging all wax possible.

Although no one would consider throwing away his cappings, it is surprising that so many overlook a less obvious but significant source of wax. This secondary wax source is the wax in burr comb, volunteer comb and other bits and pieces of wax that so often are left lying beside the hive and ultimately lost. It is quite surprising how quickly these little bits and pieces can accumulate. We are pleasantly reminded of this each year here in the breeding lab, as the small cakes of relatively clean light wax come in from the little solar wax melter. The fellows working the outyards save all scrapings and comb, then dump them into the solar extractor when they return to the lab. The same is true of the wax from the little four-frame nuclei located here at the home yard. Because many of the nucs are run one comb short to more easily facilitate queen spotting and reduce possibilities of queen damage when removing that first comb, we often find nice chunks of volunteer comb. Those that contain honey are laid in front of the nuc (if the robbing season isn't upon us) and when dry, they are also melted up.

The solar extractor mentioned above is an ideal way to handle wax melting, as it keeps wax rendered out as the season progresses. This eliminates the wax refuse accumulation that invites wax moth problems. The solar extractors are also inexpensive. Varying sizes can be purchased ready built, or they can be built simply and easily at home. One other advantage of the solar melter not mentioned so far is the lightening effect produced by gentle sun bleaching.

For small amounts of cappings, the solar melter will also do the job. The cappings may either be washed first to remove the honey adhering or the wet cappings can be placed directly in the melter. When this is done, there will be a layer of honey beneath the wax in the collecting pan. This honey may be slightly darkened, but can be readily fed back to your bees. The wax layer on top can then be washed and is ready for sale.

If a solar melter is not used and size of operation does not justify one of the several cappings melters on the market, small amounts of cappings can be melted in the kitchen, providing you have an understanding wife. However, wax is flammable, so direct heat under the wax should be applied only with extreme care. Direct heat will also tend to darken both wax and honey. Best results will be obtained if a double boiler approach is used.

After gentle heating, the liquid wax can be skimmed from the top and poured into suitable molds, or else the entire container can be allowed to cool. After the wax cake on top is solid, it can be removed, washed, and the bottom of the cake can then be scraped clean.

If the cappings have been washed before melting, and there is no honey to salvage, the cappings can best be melted with water added to the container. This reduces the likelihood of overheating and darkening the wax. However, avoid over-boiling, as "spongy" wax (water-soaked) will likely be obtained. If this should happen, the "spongy" wax can be reclaimed by gently melting without water.

Any container that comes in contact with hot wax must be carefully selected. Stainless steel is excellent, but may not be readily available. Aluminum is also good, as are enamelware and tin or tin-coated ones. Metals such as copper and brass, which give beeswax a greenish color, and monel metal, iron, and zinc, which impart a gray cast, should be avoided.

Because beeswax contracts as it solidifies, removal from molds is usually not a problem. However, if the wax is poured into the mold too hot or is allowed to solidify too quickly, it is apt to crack in the center and then adhere to the sides of the mold. Best results will be obtained by pouring when the wax is just over the solidifying point and then covering the mold with a board or cardboard. This allows slow cooling, and the wax will contract away from the sides of the mold to form a uniform, solid cake. One other pan release aid can be used. A small amount of hot water can be placed in the bottom of the mold. The wax will solidify on top of the water so there will be no contact with the bottom of the mold.

Rendering wax from slum or old combs requires additional steps and special equipment to press the sacks of material which are kept under water, and the resulting wax will not be as light in color. Old combs and slum contain propolis, pollen grains and coloring matter from pollen, and cast-off pupal cases. The wax picks up color from these materials, and it is very difficult to remove without resorting to bleaching techniques impractical on a small scale at home.

In addition to the color problem, the foreign matter present in slum and old combs gives up the wax with which it is saturated very reluctantly. The simplest way of removing the wax involves both water and pressure. By first soaking these materials in water before rendering, they appear less capable of absorbing wax during processing. The water soaked slum material should then be repeatedly squeezed under very hot water to press out as much of the wax

Fig. 7. Solar melter. Cappings and wax scraps go inside glass-covered lid. Rays of sun melt wax, and honey and wax collect separately at base of unit.

as possible. By doing the squeezing under hot water, the melted wax that is released floats to the top of the container and can be skimmed off.

REMOVING BEESWAX FROM FRAMES

Q. Is there some special way to remove wax from the frame after the contents have been cut out for chunk honey?

A. One of the best ways to remove the wax from the frames after cutting out chunk honey is to boil the frame in a lye water solution. The strength of the lye water depends upon how fast you wish to work. Strong lye water will boil off the wax in about a minute. You will have to experiment to get the proper strength. Add more lye as the solution is used. Be careful — lye is dangerous to use. If you do not care to use lye, a careful scraping with a sharp knife will remove enough of the beeswax to allow the frame to be used again.

MELTING BEESWAX

Q. How do you melt beeswax when you do not have a solar melter?

A. Fill a washtub about half full of water and use it as a double boiler with a lard can or something similar to hold the wax. Fill the lard can about one-third full of water. Place it in the tub and put a fire under the washtub, and then start adding old comb and cappings as fast as they will melt down. After you have all of the combs melted, skim off any foreign substance that may be floating on top. Then with a dipper, take the wax off the top and put it in containers or molds you have for that purpose. The slumgum will be down in the water at the bottom and most of your wax will be clean. Be very careful as beeswax will ignite easily when near fire.

SALVAGING BEESWAX

Q. I had a weak colony that the wax moth got into. As a result, the moth destroyed the entire hive, super and all. Could that wax have been boiled and saved, or was it best to destroy it?

A. The wax could have been boiled out of the combs and saved rather than destroying it. There is a fairly high return on wax from some of that material and at the current wax prices, it does count up. Generally, any bit of beeswax available is a ready candidate for saving.

BEESWAX FOR SHOW

Q. We have a Flower Show and Honey Display here in Winnipeg (Canada) and I would like to enter our honey and wax for display. Last year I lost marks on our wax because it cracked during cooling. What should be done to prevent the wax from cracking during the cooling process?

A. The beeswax, of course, should be thoroughly clean and clear. This is accomplished by melting choice capping wax in soft water with gentle boiling and careful straining of the wax until you have a clean and clear liquid. The wax should be approximately 175 degrees F. when you cast the cake. The container in which you cast the cake should have a smooth flaring side and should be heated to about the same temperature as the beeswax prior to casting the cake. This container should be placed in an insulated box preferably not much larger than the container and with a cover for the container. Carefully pour the liquid wax into the center of the container being sure not to splash any on the edges or sides of the container and then place the cover on the container so that the wax and container can cool very slowly, taking hours to do so.

Another way is to place the container in a pan surrounded by hot water, pour the cake, and then cover the container, allowing the wax to cool and congeal slowly. Do not remove from the mold until fairly cooled.

MOLDING BEESWAX

Q. About 9 years ago in a Vermont country store I purchased some 3 dimensional beeswax figures or statues, molded in a pound mold I'm sure. Now, where can I buy or make some animal molds of my own and how do I keep the warm wax from adhering to the mold? Do you think it might be possible to use molds intended for lead if they were chilled before using to help the wax from running?

A. We have molded wax figures for years and find it a fascinating hobby. There seems to be a good market for such products. It is possible to realize a good return for both your wax and labor. It requires an outlet in a store which knows values and can charge higher prices. We use rubber molds made for plaster of Paris figures. They may be obtained from most hobby stores. We keep wax from sticking by dusting the inside of the mold lightly with corn starch. Allow the wax to cool slowly and be certain that the molds are kept level.

BEESWAX FURNITURE AND FLOOR POLISH

Q. I am interested in making furniture and floor polish using beeswax.

A. We are listing below three furniture polish formulas as well as three for floor polish. We should point out, though, that all of ‘hese recipes date back to the ’30’s or earlier. Having no personal experience with any of these polishes, the judgments of their merits will have to be left up to you.

Furniture Polish

1 pint linseed oil
2 pints turpentine
1 to 2 ounces of beeswax

Combine oil and beeswax, remove from heat and add turpentine and mix. Shake well before applying.

Furniture Polish

1 gallon soft water
4 ounces soap
1 pound beeswax
2 ounces potassium carbonate

Dilute with water to suit preference.

Furniture Polish

(Paste form — best for old furniture)
½ pound beeswax
½ pound turpentine

This also makes a satisfactory floor wax by varying the amount of turpentine added.

Floor Wax

½ lb. beeswax
½ pint turpentine
½ cup alcohol

Melt wax over hot water bath. When melted, stir in turpentine and alcohol. Stir until mixture is a thick paste and pour in jars.

Floor Wax

¼ lb. beeswax
¼ lb. paraffin
½ pint turpentine
¼ pint alcohol

Melt wax and paraffin over hot water bath. When melted, stir in turpentine and alcohol. Stir until mixture is a thick paste and pour in jars.

Floor Polish

1 pint turpentine
4 oz. beeswax
3 oz. ammonia water (10% strength)
1 pt. or less water

Melt turpentine and beeswax over hot water bath. When melted, remove from heat and stir in ammonia and water until cream is cool.

BROOD CHAMBER

BROOD REARING SEASON

Q. If bees only live for six weeks or so, why doesn't brood-rearing take place all year rather than on a seasonal basis?

A. As brood-rearing takes rather large quantities of honey out of the colony stores, brood is normally only reared when the worker bees will be needed to gather the crop. When the honeyflow diminishes, the brood-rearing slows down until it stops entirely along in late October and early November. This gives a fairly good supply of young bees which will be able to live through the winter as they will have done little or no work. The queen will rest in the winter cluster until the latter part of January or early part of February when she will resume laying to supply the colony with a work force to gather the next season's honey crop.

In some tropical areas where there is a fairly constant supply of flowers in bloom, brood-rearing might continue throughout the long honeyflow. Even in such areas, there is usually some time when the queen stops laying for a few weeks of rest.

The difference in the life span of summer and winter bees is related to the amount of work done. The winter bees do very little work so they are capable of overwintering to see the colony survive into the next season. Summer bees live only six weeks or so because they literally work themselves to death. The fragile wings become too tattered to support them in flight with their loads of nectar and/or pollen and the workers die in the field.

APPEARANCE OF BROOD

Q. I'm a beginning beekeeper and have read that you can tell a lot about your colony by the appearance of the brood. This is still unclear to me. What should I be looking for and what should I be able to tell about my colony by looking at the brood?

A. It is possible to tell many things about the overall health and strength of your colony by looking at the brood but you have to know what you are looking for. An experienced beekeeper can read signs of health and disease, strength and weakness everywhere he looks as he examines his colonies but the beginning beekeeper is apt to be confused. To begin with, opening the colony to examine it causes a disruption so the bees are not acting as they normally would inside a closed colony. This means that although just watching the bees will be instructive in itself, you will not be seeing bees as they go about their duties in a normal way. The brood, however, represents both the past and the present of the colony and as it does not change by opening the colony, you can learn a great deal about your colony from examining the brood.

As brood means the eggs, larvae and pupae of the bees, there are possibilities of three different kinds of brood: worker, queen, and drone, and you should be able to tell which is which by the way the cells are constructed, and the presence of each kind of brood will also tell you something about the strength or weakness of your colony.

Fig. 8. The bees have "drawn" out the cells of this sheet of foundation and are using it for rearing young bees (brood comb). The cells at the top of the frame are being used for storage of honey and pollen.

What you hope you'll see is a good solid concentric pattern of worker cells that are recognizable by the even cappings. This means that the queen is laying in a good solid pattern, in ever-widening circles outward from the center of the comb. Some of the cells may contain eggs, some growing larvae, and some may be sealed. If the queen has missed laying in only a few cells here and there, don't worry if the overall pattern is good. If there are storage bands of pollen and honey arched across the top of the frame, that means that the bees have stored food close by to be used in feeding the young larvae. Your colony is in pretty good health.

If the pattern of brood is scattered widely across the comb, or if there are a number of cells with bullet-shaped cappings which will contain drones, you will have reason to doubt the worth of the queen. These scattered cells and the drones can tell you that the queen is failing and should be replaced with a young vigorous queen.

If you see queen cells which, when empty, look like acorn cups built outward from the surface of the comb like rooms added to a house, and which, when sealed, look like peanuts, you will know that your queen is in difficulty. Queen cells indicate one of two things: the bees are rearing a new queen to supersede or replace their present queen, or they are rearing new queens for the purpose of swarming.

Supersedure queen cells will be fewer in number, say three or four, of the same age and stage of development, larger than swarm cells because they contain more royal jelly, and will usually be constructed outward from the surface of the comb or in some depression along the side of the comb.

Swarm cells are usually more numerous, ten or more, and are of varying size and age, one or two being started each day over a period of a week or more so that the bees can give full range to their swarming impulse, and are primarily constructed along the bottom edge of the combs.

Fig. 9. a b c
a. The peanut-shaped queen cell built outward from a worker cell like a room added to a house. b. Worker cells, some capped, others ready for sealing. c. Bullet-shaped drone cells.

Fig. 10. An excellent comb of sealed brood. Very evidently the work of a superior queen. A comb of brood like this will very quickly bolster the strength of a weak colony or build up a package colony.

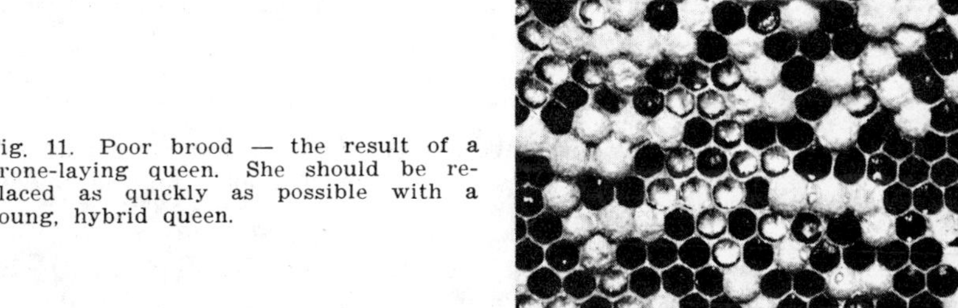

Fig. 11. Poor brood — the result of a drone-laying queen. She should be replaced as quickly as possible with a young, hybrid queen.

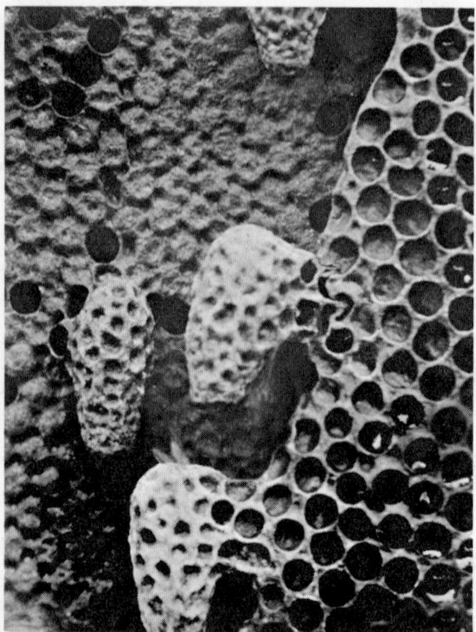

Fig. 12. Supersedure, queen cells. Notice that these cells are all of the same approximate age. Supersedure cells tend to be large and lavishly supplied with royal jelly. Notice how these cells were built next to the damaged area on the face of the comb. Some authorities tend to the belief that a queen raised under the supersedure impulse cannot be surpassed for quality. As a result of supersedure it occasionally happens that two queens, mother and daughter, inhabit the same hive.

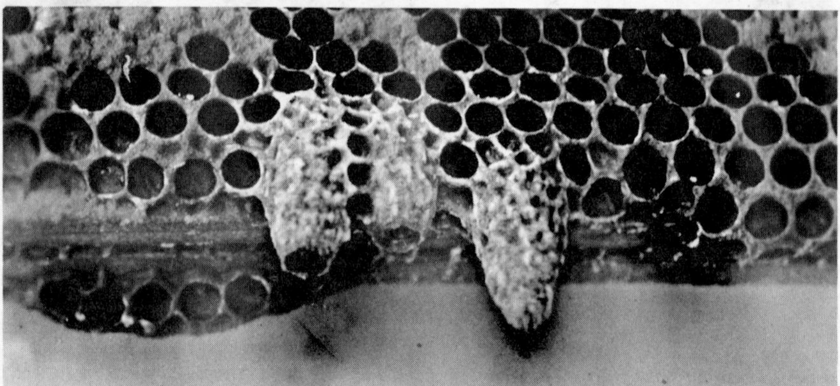

Fig. 13. Queen cells built under the swarming impulses. Notice that in this case there are three cells of varying ages. The cell on the right is already sealed and in the pupal stage. The queen cell on the extreme left is just approaching the sealing stage, while the center queen cell is still in the mid-larval stage and will be the last cell to be sealed.

MULTIPLE EGGS IN CELLS

Q. Many cells in the brood chamber of one hive contain more than one egg. Can they hatch this way or does it mean more trouble for the colony? It wasn't doing very well anyway.

A. There are several possibilities here and all of them indicate trouble within the hive. Something has to be done since it is doubtful if the situation can right itself. Since the colony wasn't doing very well, it could be that the queen is laying eggs faster than the hive bees can prepare and expand the brood nest, causing the queen to lay the second egg in some cells. If that is the case, the colony should be all right if strengthened. Check for the presence of the queen and also for any normal, capped worker cells.

It could also be that the colony has a drone-laying queen and if that is the case, the colony should be requeened.

Probably the worst possibility is that the colony has become hopelessly queenless and one or more workers have taken over the responsibility of laying eggs in the absence of the queen. These eggs, should they hatch, will only produce drones.

Since the hive is weak and you are not sure of the cause of the problem, your best procedure here would probably be to unite this colony with a queen-right colony using a sheet of newspaper. Remove the cover and inner cover from a good colony and spread a sheet of newspaper with some starter slits in it over the tops of the frames. Set the troubled colony on top so that only the newspaper separates the two colonies. The bees will gnaw through the paper and the two colonies will gradually become united.

1 OR 2 BROOD CHAMBERS

Q. I have just started keeping bees and have two hives. I am presently using two brood chambers plus a shallow super for extracting honey. I have noticed that during the honeyflow they filled the top brood chamber much more quickly than the honey super above it. Even though I had occasion to replace about three full combs from the brood chamber with foundation they still filled those more quickly than the above honey super. I have noticed, too, that the queen did not lay in the top brood chamber, but that it was filled with honey only.

What I would like to know is, is it necessary to have two brood chambers, or is there any advantage in doing so? It seems to me that the extra brood chamber could be replaced with honey supers and thus I would be able to extract that much more honey. I would very much appreciate you answering this question for me.

A. In your particular area (Bahamas), a second brood chamber would not be needed. It is a common practice in many tropical areas to use only one brood chamber.

In localities where the honeyflow is constant, the queen will usually not lay as many eggs at one time as queens do where the honeyflow is shorter and more intense. In northern areas the stimulation of the increasingly longer days causes the queen to start laying many more eggs as the days do get longer. Thus, by the time the honeyflow has started, the colony has built up to a tremendous population and is able to take advantage of the intense, but comparatively short honeyflow that occurs in the more northern areas. As a result of the extra space needed by the queen to build up larger populations of bees, many beekeepers do use a second brood chamber.

However, in areas such as yours, the honeyflow does not reach intense proportions, but is a constant honeyflow that occurs throughout the year. The queen lays just enough eggs to keep the colony built up to normal levels to take advantage of the existing honeyflow. If she lays too many eggs there will be too many bees for the amount of nectar available.

We would suggest that you extract the honey in your second brood chamber and, as you mentioned in your letter, replace it with a honey super. Then, as the bees fill up the super, you can extract it and replace the combs.

We might mention that in some countries such as Mexico, beekeepers will even go into the brood chamber and scrape the cappings off sealed honey that has been stored in the brood chamber. This condition is usually caused by the short wintertime slowdown in egg laying. As the bees move the open honey up into the supers, the queen is able to lay eggs in the empty cells. From the increase in egg laying comes stronger colonies which are able to take advantage of the next honeyflow.

DOUBLE VS SINGLE BROOD CHAMBER

Q. Is a double brood chamber better than a single brood chamber?

A. In most cases it is probably better for a colony to have two brood chambers. Most queens are capable of laying sufficient eggs to use this space. The greater the amount of brood, the greater the size of the colony will be. The larger colony has the better potential for production of a big crop, as well as having plenty of room for storage of pollen and honey for winter stores.

EQUIPMENT

Q. I would like to use (nine) frame Stoller spacers in the brood chamber. This makes for easier removing and installing of frames. Would this be okay to practice? I have heard this will stop queens in their egg-laying capability.

A. Many beekeepers do use nine combs in 10-frame brood chambers, whether or not they use Stoller spacers. This is fine as long as they are **combs.** If using frames with foundation to be drawn out into combs, it is better to use 10 in each body. This keeps the proper spacing so that the bees build the new combs on the foundation cell bases rather than making burr combs between the frames. We have not heard of the nine-frame spacing causing queens to quit laying.

CENTER TO CENTER DISTANCE IN BROOD CHAMBER

Q. What should be the distance center to the center of frames in the brood chambers?

A. The frame that is in most common use today is the standard Hoffman. The frames would have a spacing center to center of 1⅜ inches. The industry seems to have pretty well settled on this type of spacing in the brood nest.

REVERSING BROOD CHAMBERS

Q. Is it necessary or helpful to the bees or beekeeper to reverse the brood chambers around?

A. When brood rearing begins in late winter or in early spring, the brood area is usually to be found in the upper portion of the brood nest at the time of the first thorough examination of the colony. The brood area will mainly be in the top hive body when two or more bodies are used for wintering the colony, or in the super and the top of the brood combs of the hive body when the colony has been wintered in a single hive body with a shallow super used as a food chamber. Inasmuch as the tendency of the colony is to expand the brood nest upward, the queen will not readily move down to empty combs. Therefore, it is good practice to set the top hive body or super on the bottom board, setting the bottom hive body on top, thus reversing the parts. This reversal brings most of the brood to the lower position with a less amount of brood at the top. It also breaks the circle of the brood, placing the two segments in inverse positions. The queen then moves upward readily, first into the empty portions of the brood area and then into the empty combs above. At the same time, the house bees rearrange the stores of the colony around the new brood area. This stimulates the colony to greatly increase its population. With a good queen in a populous colony, it may be necessary to reverse the brood bodies again before the flow occurs.

Before the beginning of the main honeyflow, the bodies that earlier were placed on the bottom boards in the reverse position will have become empty through the emergence of the brood and the rearrangement of the stores of honey. They then should be returned to their original positions. When the flow begins, they will be used for storing the first part of the flow. At the end of the season, these bodies will be well provisioned with honey and pollen and should be left on the colonies as food reservoirs for winter. During the flow, supers needed for storage of surplus honey are added above these food chambers.

"HONEYBOUND" BROOD CHAMBERS

Q. We need some suggestions for a solution to the problem of honey bound brood chambers. Our colonies are large and occupying two brood chambers at present. They also have plenty of super room on top but are depositing honey in the brood chambers far in excess of the necessary amount needed for brood rearing, resulting in lack of laying room for the queen.

A. There is quite a genetic variation in different strains of bees and their tendencies to store honey in the brood nest. It is extremely difficult to get some bees to go above the brood nest and into the supers, while others tend to be just the opposite. This situation can be compounded considerably by the use of queen excluders. Oftentimes, bees are reluctant to operate up through the queen excluder, particularly at the beginning of the season. This can result in a honeybound brood chamber. The simple removing of the excluder does much to alleviate the problem.

One other contributing factor may be a poorly laying queen. A good vigorous young queen will usually put out so many eggs that she doesn't allow bees to store the honey in the brood nest.

You might also remember that sometimes this problem can correct itself rather quickly. The process of brood rearing is very expensive, honeywise, and entire frames of honey can disappear to be replaced by frames of brood. If you operate supers and hive bodies the same depth, the quickiest and easiest solution may be to pull some of the full frames of honey from the brood nest and place them above, replacing with empties in the brood nest. This would lure the bees to the upper supers for honey storage. If you use smaller size supers, a trick that we sometimes employ is to scratch or score the sealed honeycombs in the brood nest with something like a cappings scratcher. Once these cells are open, the bees will normally follow their natural tendency of moving honey up from the brood chamber.

CAPPINGS OF BROOD CELLS

Q. Does the capping of brood cells contain any pollen or is it made from beeswax only? I notice that the cappings on brood cells are darker in color than the cappings used to cover cells of honey. Some beekeepers think that the reason for the darker color is that the bees which cover the cells make it this way by being over it all the time. This may be true to some extent, however I think that pollen is included in this capping and that is where the young adult bee receives its first food containing pollen as it eats itself out of the cell.

A. Bees normally use a mixture of wax, usually the older wax, and chewed up pupal skins to make the fibrous brood capping permeable to air. As far as we know, pollen is not incorporated as such for providing the young bee with food.

HONEY INTO BROOD

Q. I am a novice beekeeper and I have about 18 colonies of bees. Last year on Labor day I was given 5 hives of bees and then this summer they did not make any honey at all. I did not like them at all so I raised some queens and requeened 3 of them this summer and in the spring I was given a swarm of bees, and at the same place I was given a box of bees. Well, the box I put in a good hive and the swarm in another. Now what I would like to know is this. The two hives I did not requeen and the box and swarm all started to take right off and make honey and raise brood just like a good colony. Last month when I checked them they were in pretty good shape with a good store of honey and brood. Then today I checked them and they did not have any honey at all. I mean they were empty of honey. But they had plenty of brood in all stages and they did not look like they have been robbed out at all, but like they had eaten all the honey. Could you tell me what has happened?

A. Your complaint is not uncommon this year. Many beekeepers report a very light crop of honey. During a period in which there is no honeyflow, even in the summer, the bees are forced to use the honey which they had previously stored. If your colony is headed by a good queen and there is food available, the bees will see to it that any brood in the hive will be provided for. All of your honey has gone into brood. You may find it necessary to feed the bees this fall.

REPLACING BROOD COMB

Q. Should brood comb be replaced often?

A. One will note that a brood comb that has been in use for several years will appear to have smaller cells than one recently drawn out. The reason for that is that each time an egg is placed in that cell, the bee egg hatches into a grub, then forms a sort of cocoon around it while in the transformation state and when it hatches, leaves a fine skin or shell inside the cell on the wall. Consequently the cell becomes smaller each time a bee hatches. For this reason a smaller cell might have a tendency to produce a smaller bee, which could be inferior in nectar gathering. Therefore the combs should be removed at least every ten years, if not more often.

CANDLES

CANDLE MAKING

Q. I have a small amount of beeswax that I have melted up from cappings, and as shipping costs are so high I would like to try making candles of some of this wax. My wife has been to a class in candle making and will attend at least one more class. They are not using beeswax, and I remember the grand array of candles I saw when I visited your plant. It seems to me beeswax is the best candle wax one could use. My wife thinks there should be some process of bleaching the wax to make it white. Is this advisable? Do you have any literature that gives directions or instructions for homemade candles? Or any other Home Arts for using beeswax?

A. We don't know of any good way of bleaching a small amount of wax for home usage. That is, unless you are willing to be patient and let the sun actually do the bleaching for you. Something that you can do would be to heat the wax gently over water, therefore allowing the wax to completely melt. If you can at all avoid it, the water shouldn't boil. If the water and wax boil too much together, you'll get a spongy type of wax out of it that requires some additional melting to correct the situation. By getting the wax completely melted over the water and stirring the water and wax together gently, the honey that may be adhering, as well as other incorporated particles tend to drop out leaving the wax clean and much more suitable for burning. Simply remove the wax water mixture from the heat, and let it cool down slowly. The wax will solidify on top of the water, and can be picked off quite easily. If there is any debris on the wax water interface, it can be easily scraped from the wax cake.

The chemical bleaching procedures that we use for whitening the wax are quite involved, and involve some rather caustic chemicals. They would not be suitable for home use. One thing that you could do would be shred wax into small particles, or ribbons, and expose it to the sun's rays. This is the method that was used to bleach beeswax years ago, and although it doesn't produce a white wax, it is a nice cream color if you have enough sun and are patient enough.

There has been a book about making candles written by a beekeeper's wife. The title of the book is Wick, Wax and Talk by Mrs. Owen Wilhelm. By writing to Mrs. Wilhelm, Wilhelm Honey Farm, Erick, Okla. 73645, you can get further information on the book.

Now that the more serious work of harvesting the honey crop is over, it is time to start thinking of a few of those lighter, more fun-type projects. One of these projects would have to include making your own beeswax candles. Such an undertaking may seem a bit difficult, but this need not be the case.

It's really not much different from making other kinds of candles at home; a do-it-yourself project that is becoming increasingly popular. A look at the Christmas catalogs from the major mail order houses will give an indication of the scope of the general interest in home candle making. But, no synthetic-wax candle can come close to capturing the magic of one of pure beeswax. The soft rich beeswax color, combined with the beeswax aroma and burning qualities, puts beeswax candles in a class all to themselves.

The first step in any such project is that of securing the proper materials and utensils. Finding the basic raw material, the wax, is no problem. Simply choose a cake of wax that is colored to please you. Securing the proper wicking can be a bit more troublesome. Common kitchen string or cord should not be used — you will need to go to a hobby shop or other supplier who handles candles supplies for genuine wicking. They can help you select the size wick needed, depending on the size of candle you intend to make. The rest of the equipment needed can be found in any kitchen, providing permission to use those facilities is forthcoming. Regardless of the type of candle to be made, you will need a double boiler for heating and cleaning the wax, a kitchen thermometer for obtaining proper wax temperatures and a suitable cup for pouring wax. For hand-dipped tapers, a tall narrow container such as a juice can will be needed, as well as a larger container with heated water to help maintain the proper wax temperature. For molded candles, the proper molds need to be secured.

First, prepare the wax. Select a cake of clean wax of the color desired. If a cleaning step is required, gently melt the wax, with water added, over the double boiler. Be careful not to allow the water to boil, as a water-wax emulsion can result. Rather, melt the wax over the water, then stir the mixture, after which it should be allowed to settle. Normally, the water will settle out to the bottom carrying with it dirt and debris. The cleaned wax can be dipped off the top.

In dipping tapers, cut off a length of wicking longer than the candle desired. Fasten a weight to one end of the wick. This will allow you to get the wick to the bottom of the wax-filled container rapidly, and will also aid in keeping the candle straight. Fill the dipping container with wax, and maintain the wax at a temperature of about 170°F. Then, do repeated dippings until the proper diameter of the candle has been obtained.

For pouring molded or cast candles, the secret again is in the temperature of the wax as it is poured. Pouring too hot and cooling too fast can cause the cake to stick to the mold and open up cracks in the middle, while pouring too cool will result in air bubbles becoming incorporated. Again, the 170° temperature will not be too far from ideal. Once poured, cover the mold and allow to cool very slowly.

Colors can easily be added for either type of candle. However, be sure to use oil-soluble dyes. Coloring pigments tend to clog wicks and water-soluble colors just will not mix with the wax. For a start, melted color crayons can be used. In either type of candle, the dye can be mixed with the wax first to produce a candle of solid color or it can be used only as an outer dip coat.

Now, you're all set for not only home decorations, but also have an ideal gift for friends during the holiday season. What more appropriate personalized gift than a pair of beeswax candles and a jar of honey!

CHEMICALS

CHLORDANE SPRAY IN THE HONEY HOUSE

Q. I want to know if I use chlordane spray under the hive with bees inside will I kill the bees? I sprayed around the floor next to the wall inside the honey house with chlordane. Now people tell me the bee supplies I have in there are saturated and will kill the bees. The supplies are four feet above the floor.

A. Chlordane is generally a bad pesticide to use around bees. We would not recommend that you spray under the hive with the bees inside nor that you use it to spray in your honey house. The fact that your equipment is four feet above the floor will probably mean that it is all right. However, we certainly wouldn't recommend that you use the spray again in the future. It is true that beeswax will absorb the chlordane and then kill bees when you use the combs on the bees again. Many of the sprays used around beehives are not really necessary if the colonies are kept in good condition. The bees can defend themselves rather well if they have a strong colony. Likewise, many honey house sprays are unnecessary if the building is kept clean, tight, and dry.

CHEMICALS

Q. I have read that beekeepers use chemicals to drive the bees off combs of honey. Could this result in harm to people when the honey is eaten?

A. The chemicals used as repellents, if properly used and the supers thoroughly aired, do not contaminate the honey through absorption and, therefore, the honey would not be harmful to people when it was eaten. However, the bee blowers are now being used extensively and anyone who has blown bees from supers would not go back to the use of chemical repellents. So, we hope the use of chemicals will ultimately disappear from beekeeping practice.

COMBS

Q. I have some empty combs that I would like to treat and store until next spring by treating them with paradichlorobenzene. Next spring I plan to use these same combs for gathering my surplus honey which I will later extract and sell. Is there any danger of contaminating the honey in any way by doing this? Also what do you recommend as the best way — in my section of the country — to store empty combs and to protect them against wax worms?

A. Paradichlorobenzene is one of the standard materials for treating combs to prevent wax moth damage. We would suggest that you air out the combs well before putting them on the bees. It should have no effect on the extracted honey you produce.

Stack the supers on a level floor with about a tablespoon of paradichlorobenzene on a piece of paper on top of the frames of every fourth super and cover the stack. Put a queen excluder under the pile and on top to prevent mice from entering and damaging the comb.

CHEMICALS

Q. I am interested in a clipping I have that states that diluted acetic acid drives the bees upward out of the hive. Do you have any information on this or could you refer me to anyone else that would know?

A. We're not familiar with anyone who successfully uses acetic acid to drive bees out of the hive. We do know that work some years ago was done showing that under some conditions, acetic acid did drive bees downward. However, the fumes were retained by both honey and wax. As a result, its use cannot be recommended.

There are a number of other repellent chemicals that are used with much more regularity by our industry. The two that are most commonly used now in removing honey supers are benzaldehyde and Bee-go (another name for butyric anhydride). Carbolic acid was used for many many years until the Food and Drug Administration felt that this could result in the addition of phenols to honey. Its use is no longer recognized nor allowed, but it was a very effective bee repellent.

Many beekeepers in the last couple of years have been making the change from repellent chemicals to the use of bee blowers for removing honey. Instead of driving the bees out of the honey supers with chemicals, they just blow them out with forced air.

Fig. 14. Bee blower, complete with gas-powered engine, plastic hose and spout.

COLONY LOCATIONS

NUMBER OF HIVES IN ONE OUTYARD

Q. I would like to know how many hives of bees to keep in one place (Tennessee)?

A. The number of beehives which may be kept in one outyard varies considerably depending upon the number and kinds of honey plants which are available for the bees to work on in the area. A number of years ago we kept bees in southwestern Iowa and most of our yards had from 80 to 100 hives of bees. This was due to the fact that there was lots of sweet clover in bloom every year. In Pennsylvania, perhaps 20 to 40 colonies make a yard; while in Florida or in areas where there are honeyflows from specific crops, the number of colonies can be increased. I do not think one hundred colonies would be too many in the citrus groves. On the other hand, in this particular part of west central Illinois, it is preferable to have no more than 25 hives in each location.

Since the bees from a single colony will range over an area of approximately 12 square miles, that is a circle with a 2 mile radius, this makes it understandable that there can be a lot of honey plants in the area.

COLONY LOCATIONS

Q. How close should one apiary be placed to another (Wisconsin)?

A. Over the years, there has been a lot of controversy and discussion concerning the placement of apiaries. Some states have regulations which specify how close registered apiaries can be placed. South Dakota is an example; they require apiaries to be at least 3 miles apart. We feel certain that this large a distance is not necessary. It is quite likely that any floral source at a distance of over ½ to ¾ mile away from the hive would be too far away for the colony to take good advantage of the nectar flow. Certainly, bees can and do fly farther, but it is doubtful whether their trips would be worthwhile. Therefore, if you wish to place your apiaries as close as a mile apart, there is good evidence that it would be quite all right to do so.

Here in Illinois the registration of apiary locations and names of all beeyards, with locations and number of colonies in each are required. Also a sign must be placed in a conspicuous place giving permit number, owner's name, address, and telephone number.

Registration is becoming of more importance as losses from insecticide spraying programs are developed and increased. Knowing what beeyards are in a given area allows the insecticide operators to warn the beekeepers. They can then cover the colonies, move them, or whatever, to avoid a spray loss.

COLONY LOCATIONS

Q. I have a problem early in the spring before the bees have anything to forage on. I have about 50 hives located 200 yards from a restaurant and service station. The bees hang around this station annoying the owner and customers. Is there any spray or any other methods I could use to discourage them without harm to the bees?

A. The bees that you have located close to the restaurant and service station should not be bothering the rest stop unless they have some specific need for being up there. The most logical explanation would be that the bees are after a source of water. You might prevent it by providing a source of water right among the colonies, thereby eliminating the problem. It would be most desirable if this could be done in early spring before the bees establish a pattern of going up to the station for their water.

BEE PASTURE

Q. I have an opportunity to place colonies on a dairy farm, near Hagerstown, Maryland, that grows alfalfa and clover. In order to maximize the protein content of the feed the farmer cuts the alfalfa between late bud and early bloom. This generally occurs between June 1 through 10 and is followed by two more cuttings during the summer.

How long would he have to delay cutting the alfalfa in order for me to get a honey crop? Do you recommend a management scheme we can work out to optimize nectar flow and protein content? The farmer is not interested in raising seed crops.

A. Almost all of the agricultural universities and colleges around the country now recommend that alfalfa be cut in the pink bud stage. In that stage the farmer gets a maximum amount of protein in his hay. Actual protein levels may vary between 17 to 21 per cent. If the farmer waits until the alfalfa actually blooms, then the protein level can sink to as low as 12 to 14 per cent which is a considerable loss to him. There doesn't seem to be any kind of a management scheme that can both optimize the nectar flow and protein content.

One thing to remember, there are seasons when it is so wet that the farmer can't get into the field. Under these conditions, the alfalfa does go ahead and bloom and you should be able to count on some rather good nectar days before the farmer actually gets into the field.

Still another condition sometimes exists. In your area most farmers try to take three cuttings of alfalfa per year. They may make the first cutting and the second, but frequently they do not get the third cutting because it has been too dry to have sufficient alfalfa growth for hay purposes. In spite of this, and in spite of the fact that the alfalfa is rather low, it continues to bloom. Your bees should, under such conditions, get considerable nectar in the late summer as a result of the above condition.

HIVES AND STOCK CATTLE

Q. Is it possible to keep beehives in a pasture where the cattle are being pastured.

A. We have had hives located in a pasture where cattle have been kept. Normally with old stock there is no problem. However, the chances of having some hives tipped around are increased with young stock. You are also likely to find a number of hives shifted on their bottom boards where the cattle have tried to scratch themselves or rub against the hives. All in all, we have found it the most convenient to put up a fence around the hives to keep the stock at a short distance.

COLONY MANAGEMENT

Gladstone H. Cale, Sr. was editor of the *American Bee Journal* from 1928-1965 and had many years of experience in honey production in the apiaries of Dadant & Sons, Inc. and as the operator of his own apiaries. During that period of time he answered thousands of questions concerning all phases of beekeeping and his remarks on management techniques are as pertinent today as they were when he refined the techniques through long hard hours in the apiary:

Success in beekeeping depends upon a proper exercise of the knowledge of colony organization, growth, and behavior in relation to environment as affected by seasonal changes and the occurrence of nectar- and pollen-bearing flora. In beekeeping it is not possible to use fixed rules or exact routines. No two seasons are ever alike and the beekeeper who has the truest understanding of the habits and activities of bees and of the fundamental reaction of the colony to its environment is the one who is most likely to succeed.

In the intelligent management of a colony of bees there are practices which do not come in any natural succession throughout the season, but may be necessary to employ whenever conditions require their use. Each of these practices is important and vital to the well-being of the colony and to the production of a maximum crop.

The major portion of the total honey crop of the United States is harvested from May to September, although in the South and in California, crops are produced as early as March and April. It is likely that at least half of the total crop of the country is produced before July 1. It follows, therefore, that the inactive, preflow, honeyflow, and afterflow periods may be occurring at about the same time in different parts of the country. Although many of the answers to beekeeping questions are based on beekeeping in the Midwest, which seems to represent an overall average, beekeepers in other areas will be able to apply the same procedures, timed to their conditions.

The honey-bee colony must come through winter and into early spring with enough of the winter population of worker bees remaining to support the queen in her egg laying and to care for the brood, so that the colony will increase in population until it reaches a high point at the beginning of the main honeyflow. This is a matter of timing and proper late summer and fall management has much to do with it so a colony can both adapt successfully to winter conditions and also be productive the next season.

When the food reserves of honey and pollen are sufficient for winter, many colonies with capable and efficient queens will begin brood rearing in the latter part of the winter, and will add enough young bees to replace their winter population in early spring. Colonies with insufficient stores or with less satisfactory queens will have little, if any, winter brood. They may not start brood rearing until spring is at hand, and often they are unable to build to peak strength by flow time.

RECORD KEEPING

Q. Being a newcomer to beekeeping, I am still in the "extensive reading" stage of this most fascinating of hobbies. I have frequently read of the importance of record keeping in connection with hive management but have not done so with my few hives. I do plan to increase the size of my apiaries in the spring and would like to know more specifically what should be observed and recorded for ultimate management. I feel reasonably sure that experience has dictated a need for certain records to be kept and wonder if there is some specific record book I should use.

A. It is true that well kept records can make it much easier for one to plan future work by knowing what has been done in the past, but as beekeeping is pretty much of an individual operation, no one record book seems to fit the job for all people.

The most successful record books that we know of are the simply written logs or journals of such master beekeepers as the Rev. Langstroth, Charles and C. P. Dadant etc. They seemed to prefer sturdily bound journals in which they could simply note down observations and later act on those observations. Regardless of what system of records you keep, it should be considered as a tool, something to help you improve your management. Records just for the sake of keeping records is a waste.

There are a number of things you might do. Number your hives and keep a log of observations — "No. 2 hive propolizes excessively.", "No. 18 is working quietly.", etc. Let these observations be memory joggers to stir you into finding out why these conditions exist and change them to suit your plan of honey production. For example, you might note the condition of the queen, her age, whether she's laying well or poorly; when the hive was requeened; whether she's marked or not, and if so, what color, and what kind of a queen is she. Another item for observation concerns the amount of food in a hive at a given time including both honey and pollen. The strength of each colony is also important, whether it's weak and needing additional bees or brood, or whether it is too strong and something is needed to prevent swarming. Another worthwhile item, too, is the amount of honey produced by each colony. If the pages in your log are marked off into columns for each of the above or for whatever phases of beekeeping you want to record, you will be able to compare your hives and their conditions at a glance down the page.

Records written in books are not the only kind which may be useful to beekeepers. Many use a system of placing rocks, for example, to indicate colony conditions. These rocks are placed on the hive cover in certain ways to indicate certain conditions. Just before leaving the yard a summary of these conditions can be made so that the proper equipment is brought back on the next visit. Others may make notes on the sides or tops using carpenter crayons, or such. These are more permanent than the rocks on top which may get knocked off or moved but a disadvantage of placing marks on the hive cover or sides of the hive is that such marks are meaningless if the equipment is interchanged with that of other hives.

The systems are as varied as beekeepers — the most important thing to remember about whatever system of record keeping you use is to act on it and make it work for you to become a better beekeeper.

COLONY POPULATION

Q. Can you give me the approximate number of bees I would want in a medium strong colony?

A. The answer is going to vary considerably as the population of a colony of bees fluctuates according to the season of the year and as a beekeeper, you will want the population to vary as this is one of the factors in manipulating the colony to produce a surplus of honey. If your colony reached and maintained a peak strength all year, they would gather a surplus of honey and then consume it, leaving you with nothing for the time and expense.

It has been estimated that there are approximately 3500 bees to a pound. Installing a 2 pound package of bees in early spring would give 7000 bees plus one queen. That colony can be expected to dwindle somewhat as some bees will die before the first of the new brood begins to emerge. A spring colony might be considered strong if it contained three to four pounds of bees but the same colony would be expected to have ten or more pounds of bees two months later if it was to remain a strong colony. That same colony might well reach a peak population of sixty to eighty thousand bees and might well reach a hundred thousand bees, depending upon the strength of the honeyflow and the egg laying capacity of the queen. Such a peak population would be maintained for only a short time to allow the workers to take advantage of the honeyflow and then the amount of brood-rearing would be reduced as the honeyflow dropped off. Older bees die off as they become exhausted with the field work until the colony decreases to a size that can form a winter cluster and survive until the next season.

The bees regulate their numbers according to the needs of the season as it is their instinct to store a surplus of honey during the honeyflow and reduce the amount of honey consumed at other times during the year.

COLONY MANAGEMENT

Q. I recently purchased five hives of bees. They are three hive bodies high and ten frames in each body. In examining these I was surprised to find brood and sealed honey in the top one (weighing about 80 pounds) and also brood and sealed honey in the second.

My questions are: should I remove one frame in each super to allow them more room? and should I put a queen excluder between the third and second body and requeen? This is my first year with bees.

A. It sounds as if your five hives contain good, strong colonies. We would suggest that you do not put an excluder between the second and third bodies and requeen. It is true that many beekeepers do use only nine combs in a ten-frame body. (When using frames with foundation in them, there should be ten per body.) We would like to suggest the following possibilities for you.

If you do wish to remove one comb from each body, you could take good combs of brood and also food combs, with the adhering bees (but not the queen), and make divides, or new colonies. The combs of brood used for this should be mostly sealed brood which has bees starting to emerge. In preparation for this, the queens should be ordered previously. As soon as the divides are finished, they should be moved to another location a mile or so away. This is to prevent the field bees from returning to their original locations, thus leaving the divides with too few bees to care for the brood properly. The divides should be requeened the same day.

Since the two top bodies have brood in them, we assume that the bottom body does not. You might consider "reversing" each colony. Set the empty body aside. Put the middle body on the bottom board and the top body on top of it. This will give the colony room to cluster in, and also to expand the colony upwards. If you wish to keep the queen from laying in this top body, the excluder can be placed between the two top bodies.

CROSS COLONIES

Q. I have 10 hives of bees and have kept bees for several years, but this is the first time in late June and early part of July that I have seen them so hostile and ill-tempered that they almost eat me up when taking off honey. Lately, they have even followed me to the house and run me in. Please advise your experience in this case.

A. We would suggest requeening the bad-tempered colonies if you can determine which ones are doing the stinging. If all of the colonies seem to be equally bad-tempered, it might be wise to check over the surrounding area carefully to see if something is keeping them aroused. If possible, locate the colonies where a natural screen of some sort separates them from the house.

COLONY MANAGEMENT

Q. Do beekeepers wear white because bees recognize this color?

A. It is our opinion that the statement "that beekeepers wear white clothing because bees recognize this color" is not correctly stated.

It is true that beekeepers wear white clothing to enable them to work the colonies of honey bees without being stung to any extent. This is because the compound eye of the honey bee recognizes the movement of a dark object much more quickly than one which is light colored or white. This movement of a dark object excites the bees to protect themselves and this often results in stinging. For example, a beekeeper working bees will often remove a wrist watch because angry bees are inclined to head toward it. In a similar way, they will fly toward the eye or the dark opening of a mouth in much a similar way. This is why beekeepers wear bee veils. Similarly a beekeeper wearing white socks will seldom be stung around the ankles but if he wears black socks, he may be stung numerous times there.

The vision of the honey bee is quite different from that of a man. The colors red and yellow at that end of the spectrum are seen as greys and even black objects and, therefore, the bee is color blind with respect to these colors. The bee does see the greens, the blues and the violet colors and then continues to see in the ultraviolet range where man does not see. Colin G. Butler, in his book entitled "The World of the Honeybee" states "Some white objects absorb ultra-violet light while others reflect it. This means, of course, that two objects which appear equally white to us may appear as of two quite distinct colors to bees — the first can readily be distinguished by them probably as a shade of blue-green, whereas the other kind of white object probably appears as a shade of white. He continues, "Therefore, since most white flowers absorb ultra-violet light, such flowers probably appear to be blue-green to bees."

We do not know whether the white cloth of a uniform for beekeepers would reflect ultra-violet light or not but have been informed that ultra-violet light is used to determine the whiteness of clothing in studies of various detergents used for cleaning them. It is claimed that the use of ultra-violet light

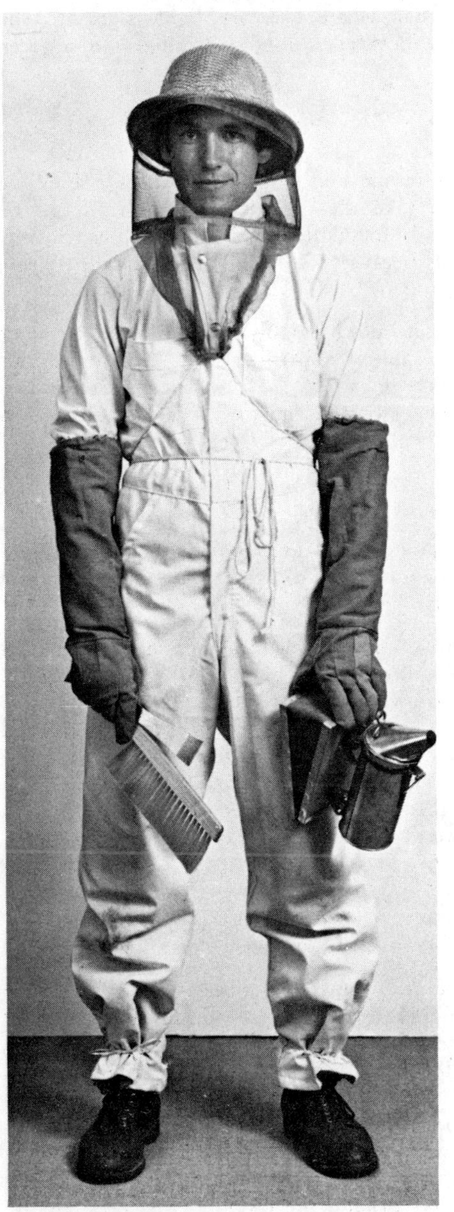

Fig. 15.

This beginning beekeeper will be able to manipulate and observe his hives at leisure, knowing that he is protected against stings and well-equipped to handle any situation.

His white coveralls, tied at the ankles, protect him and yet allow maximum freedom of movement.

The woven fiber helmet is cool and comfortable and supports the wire and cloth mesh veil away from the face.

The full, elbow-length sleeves on the gloves offer maximum protection. Although the experienced beekeeper may discard the gloves, they are helpful to the beginner until he becomes accustomed to an occasional sting.

The smoker will be used to produce a calming effect on the bees. When smoke is applied, the bees gorge themselves with honey and are then easily handled.

The hive tool is an all-purpose tool for nail-pulling, prying, scraping burr-comb and propolis and indispensible in handling the colony.

The soft bristles of the bee brush will be used to gently brush the bees from clothing, hive parts, or frames of brood or honey.

enables one to tell whiteness better than where ordinary light is used. Who knows perhaps the white uniform of a beekeeper might also appear blue-green to bees.

COLONY MANAGEMENT

Q. I have a hive that was given to me and the brood frames are falling apart from age. I have been told to take the old hive body, turn it upside down on top of a new hive body with foundation and, as the bees emerge, the queen will not lay eggs in the cells because of their downward slant. Please advise?

A. Turning the old hive body upside down might well work and it might be interesting to give it a try for the sake of experimenting. A better suggestion would be to remove the hive cover, set a new body with combs or frames and foundation on top of the old hive body and replace the covers. The bees will work up into the new part and then the old body can be removed.

COLONY MANAGEMENT

Q. My present hive has three frames which are almost solidly connected with comb (they weren't put in straight). Should I break up the comb and put in new foundation or leave it as is?

A. Move the three crooked combs to the outside positions in the hive and prevent the queen from laying in them if this is at all possible. When there is a nectar flow in June and bees are drawing comb readily, replace them with frames containing straight, full sheets of comb foundation. If the crooked combs contain honey, place them in a body on top of the colony and above the inner cover with the hole open. The bees soon will take the honey down and dry the crooked combs. They then can be removed and the combs cut out of the frames and melted into beeswax.

SHARE-CROPPING ARRANGEMENTS

Q. What is customary in a share-crop bee and honey operation?

A. There have been a number of arrangements for bees in the past, none of which seem to be particularly applicable to the present day situation. In the past it was normal for the owner to supply the bees and the extracting equipment with the person running the bees then supplying all the labor plus the truck and truck operation costs. The one running the bees also supplied the queen and sugar with the crop being split 50-50. Under present day labor costs, this would seem a very unfair arrangement for the person actually doing the bee operation. Some variations of the above have been called to our attention and most of these variations revolve around the owner of the bees supplying half the sugar and half the queens.

We would tend to look at the operation in light of the rising costs of labor. If the operator is to supply the trucking and all of the sugar costs and queen costs plus the labor, then the operator should certainly have no less than 60% of the honey produced and even 70% might be a more fair arrangement.

COLONY MANAGEMENT

Q. I could use some tips on colony management in general. I guess the problem is that I am not as observant as I should be but I am always playing "catch-up ball" with my bees and trying to correct a situation that is nearly out-of-control before I notice what is going on.

A. We wish we had an instant answer for you but experience is going to have to be the teacher in this instance. The best we can do for you is to advise that when you are going through your colonies, try to think of what the bees are doing to cause the end result that you are viewing. In talking with other beekeepers, it has always been interesting to us to notice that some only look — direct their eyes — at their colonies, while others see — and really see — what they are looking at. One man may lift out a comb and note only that it is a comb of brood, while another could lift out the same comb of brood and see the sunken, slightly perforated capping on one cell which might be both an indicator of and a concealment for a diseased pupa. Or perhaps there might be a number of protruding cells — drones being reared in worker cells, which could mean a failing queen.

This ability to see through surface conditions helps develop the ability to evaluate the actions of the bees, the supply of honey and pollen, the relative number of drones, the presence or absence of queen cells, and so on through the total of variability which is a colony. The beekeeper who sees, truly knows the condition of his hive of bees. To us, it is this which separates the "beekeepers" from those who "keep bees."

Many observant beekeepers can tell much of what is going on inside the hive by noticing the activity at the entrance. Should you see a number of white larvae or pupae there, or being dragged outside the hive, you can almost bet that the colony is starving and needs immediate attention.

Another possibility is, if there has been a good warm spell followed by a sudden cold snap, that the cluster may have over-expanded and then was unable to care for all its brood during the cold period. So the bees may be removing just that brood which was chill-killed when they had to contract the cluster to maintain brood rearing temperatures in the center of the brood nest.

Another tip here would be to obtain a journal and number each of your hives. When you examine a hive, enter the date, hive number, and your observations while you are still at the hive. Put down your hunches, what you think is going to happen, and later, compare the results with your journal.

SPRING

SPRING MANAGEMENT

Also see:

Package Bees
Feeding
Uniting Colonies
Brood

EARLY SPRING MANAGEMENT

Q. There are so many things to do in the spring. Is there any preferred sequence to spring management?

A. It seems as though everything is important in the spring. Much of the success of the colony during the coming season will depend upon the start the bees get in the early spring. A colony that gets off to an early spring build-up will be in peak condition to take full advantage of the honeyflow, while a troubled or weak colony will often have difficulties throughout the season.

We are going to prescribe a sequence of events for spring management but keep in mind that bees often fail to follow the same program as the beekeeper. Be alert to the needs of the bees and be prepared to deal with problems as they become apparent.

First of all, to find out where you stand, inspect the colonies. Colonies that have been wrapped with tar paper or other similar material and those that have been lightly packed can have their protective coverings removed in early spring. If the colonies were heavily packed, the packing should be left on until the season has advanced to a daily maximum temperature of about 60° F. These colonies tend to expand their brood area over a greater comb area than those with light packing and removing heavy packing early will cause the bees difficulty in keeping such a brood area warm enough to prevent loss of brood through chilling. Heavily packed colonies should be amply provisioned with stores of honey and pollen in the fall so that it will not be necessary to remove the packing before warm weather occurs.

Give all of the colonies a quick examination in the early spring and work quickly and with care, stopping if robbing becomes apparent. It is not necessary to wait for the bees to fly freely every day but don't keep the hive open too long as the brood may become chilled. If the colonies are in two or more full-depth bodies or if they are in one hive body with a shallow food chamber, pry up the top body and tip it back without removing the cover on top. Look closely among the combs to see if there is brood present. If there is no brood, the colony is most likely queenless and if the queen has failed or has become a drone layer, it will be evident by the pattern of the brood or the predominance of drone brood.

If the colony has a strong cluster of bees and a normal amount of worker brood, it will need little attention unless it is light in stores. Lift at the back of the hive and if the hive feels light, open the colony and examine the combs for stores of honey and pollen. If lacking in honey stores, the colony will need to be fed sugar syrup or given combs of honey. If there is little stored pollen, the colony should be fed a pollen substitute or pollen supplement, especially if the early spring sources are not yet available or the weather is still too unsettled for daily flights.

During this first examination, you should also be alert for symptoms of disease, especially Nosema and foulbrood. For further information concerning the symptoms and treatment, refer to the Disease section of this book. Fumidil-B® is the recommended treatment for Nosema and can be easily fed mixed in a 10-pound pail of sugar syrup (a rounded teaspoon of Fumidil-B® per gallon of syrup.) A preventive medicine for both American and European Foulbrood is a quarter teaspoonful of Terramycin®* (TM-25 Animal Formula)

Fig. 16. Properly protected — note veil, coveralls, and bee gloves — the beekeeper approaches colony from the side. Use the smoker to puff a small amount of smoke in entrance.

Fig. 17. Remove outer cover, puffing a small amount of smoke into bee escape hole in inner cover. Then pry up inner cover with hive tool, smoking under cover as it is raised.

Fig. 18. After blowing smoke gently over the top of the super, use the hive tool to pry apart the super from the hive body. Smoke gently as super is raised.

Fig. 19. The top of the brood nest is now exposed. Smoke gently and then use hive tool to pry out the outside frame. Use smoke sparingly. Too much excites the bees.

Fig. 20. The outside frame is removed and placed beside hive. This allows other frames to be examined more freely. Note how hive tool is held to allow full use of fingers.

Fig. 21. The center frame is removed from hive for examination. Note brood and quietly working bees. With judicious use of smoke, bees proceed with their normal activities.

and ¾ teaspoonful of powdered sugar, dusted over the tops of brood frames at three different intervals before the honeyflow.

Later, during the early bloom of dandelion or fruit trees, when the colony is being provided with nectar from these sources, you will be able to open the colony and make a more leisurely and thorough examination. Carefully examine the brood nest and watch for these three things: that the queen in each colony is satisfactory, that food is present in sufficient quantity, and that there is no disease in the brood.

As you finish the examination of each colony, make sure that the entrance is clear of obstructions so that the bees will have free flight, and reduce all entrances so that the colonies will be able to guard against robber bees from other colonies. Later, when the nectar is coming in heavily and the colonies are strong enough to defend their stores of honey, the entrances can be enlarged. Clean up the bee yard as you go. Start a system of simple records for each hive, noting the condition of each hive, level the hives if the winter freezing has tipped them out of position, and provide a source of water for the bees.

---

* Terramycin is a Trademarked product of Pfizer Inc., U.S.A. Both Terramycin and Fumidil-B are available through Dadant & Sons, Inc., Hamilton, Illinois 62341.

EXAMINING HIVES IN SPRING

Q. When is it safe to open hives for the first examination in the spring?

A. If the temperature is up to 50 degrees F., and there is little or no wind blowing, brood may be hastily examined, but should not be exposed for more than a few minutes. Just raise a frame enough to see brood and whether or not there appears to be any disease, then put the frame back into position, examine for stores, and then close the hive. If the temperature is around 65 degrees F., then one could go down into the hive body and take out frames and examine them all. This is not usually necessary. Just find some brood, stores, and a good sized cluster of bees and you will know your colony is in good condition. It is not a good idea to open the hive too often until the middle of May. A beginner may want to look into the hive every week during the summer. This might reduce the honey production some, but not to a great extent.

FORCING THE QUEEN TO THE BOTTOM CHAMBER

Q. I have 10 colonies and plan on building up my apiary to about 100 colonies in the next two years. Last year I did not have an extractor and rather than waste my combs, I left some of my supers on the stand which gives me several four-story hives.

Should I force the bees down into two hive bodies, remove the two top supers and replace as needed or leave the hives as they are until they are refilled? I do not use queen excluders so will the queen move down into the brood chamber when the honeyflow starts?

A. If you do not intend to use an excluder then you must practice management procedures which will keep the bees down below. You can move the brood combs down into the lower chamber or if they are in one brood chamber already, you can go ahead and reverse it and put it below where you want the bees. If you do have a good heavy honeyflow, the bees will fill up those two supers rapidly. However, if it is a poor flow, they will store small amounts in each super which will make it much harder for you to see when one is full.

If you are willing to wait until spring, the bees will force the queen down when the honeyflow starts. If you wish to have her in the bottom brood chamber sooner, you can take some of the brood combs out and put them below and she will move down. Then you can put a queen excluder on to keep her from moving up into the supers. We would suggest that letting the bees move your queen down might be the best solution.

All of this is going to be extra work and since you plan on enlarging the size of your operation, go ahead with the purchase of the extractor so that supers can be emptied to be given the bees as they need them.

SLOW SPRING BUILD-UP

Q. I have been an amateur for 17 years but here is a problem that has me puzzled. Last year I received a queen for requeening from a commercial firm. The colony came through the winter in good condition but seemed slow in spring build-up. When a second hive body seemed nearly full in early May I added a third super with foundation for chunk honey but the colony just did not increase any more. On June 25th, although there was plenty of brood in both bodies and the same queen was present, there was scarcely two supers with bees. The weather has been quite cold.

A. We wish there was an easy answer as to why a colony of bees did not store more honey. Several factors which may be responsible are:

1. Cold weather. Flowers probably do not secrete as much nectar and bees certainly do not gather as much nectar during cold weather.
2. Disease, such as Nosema, affects the adult bee population.
3. Queens. It is not necessarily true that a young queen is always a good queen. Check her brood pattern. It should be uniform.
4. Not all strains of bees are industrious, just the same as people.
5. Hive population. It takes a lot of bees to gather a crop of honey.
6. Insecticides may be poisoning the bees to keep down the field force.

Any of the above factors or a combination of them may be causing your trouble.

ADDING BROOD OR BEES TO WEAK COLONIES

Q. I have some weak colonies that I would like to strengthen either by adding frames of young brood or by adding young bees. One beekeeper friend told me he merely shakes the young bees in front of the hive and lets them crawl in and add to the strength of the colony. Which method is best?

A. The safest and probably the best method to use would be to give the weak colony a frame of sealed brood from a strong colony. The brood will soon emerge and add to the colony strength. Shaking young bees in front of

Fig. 22. A good comb of brood that could be used to strengthen a weak colony.

the hive could be used if done early in the season when there is nectar coming into the hive. Young bees are generally accepted in a different colony almost any time of the year except when robbing is going on. When the bees are shaken in front of the colony, generally the old bees fly back to their original hive and the young bees move into the new colony. You might also want to consider switching locations with a strong colony to let the weak colony pick up drifters and the returning foragers.

REVERSING COLONIES

Q. Last summer I put a queen excluder between a brood chamber and two supers. When I wanted to extract my first honey, the supers were full of brood. Now that it's spring I should like to get the colony back into the brood chamber. What is the best procedure?

A. It sounds as though the queen was either in the supers when you placed the queen excluder on the colony or the excluder may have been damaged slightly so there was room for the queen to move up into the supers. In either case, on examining your colony this spring, you may find that the queen has gone back down into the bottom, that is, if you removed the queen excluder last fall. If she has not gone back down, reverse the two hive bodies and put the upper one on the bottom of the hive so that your arrangement is an upside down hive: super, super, brood chamber. The queen could also be picked up very carefully by the wings and placed in the lower body. Reversing the colony can be done when the weather warms up. If the hive bodies are reversed, you can do this as soon as the weather will permit you to work with the bees. If the queen is moved down and the brood is left in the top hive body, you should wait until the weather has become settled, and no more cold weather is expected.

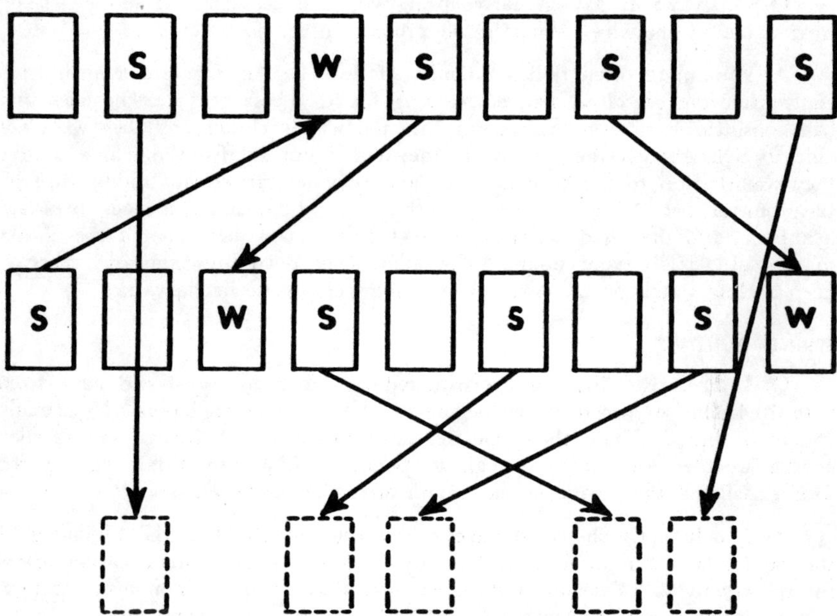

Fig. 23. Diagram of relocation. Strong colonies are indicated by the letter S; weak colonies by the letter W. Arrow indicates possible relocation.

Fig. 24. Diagram of reversal of a colony in one hive body and a food chamber. In spring the food chamber is placed under the hive body and reversed to the position at right before the honeyflow.

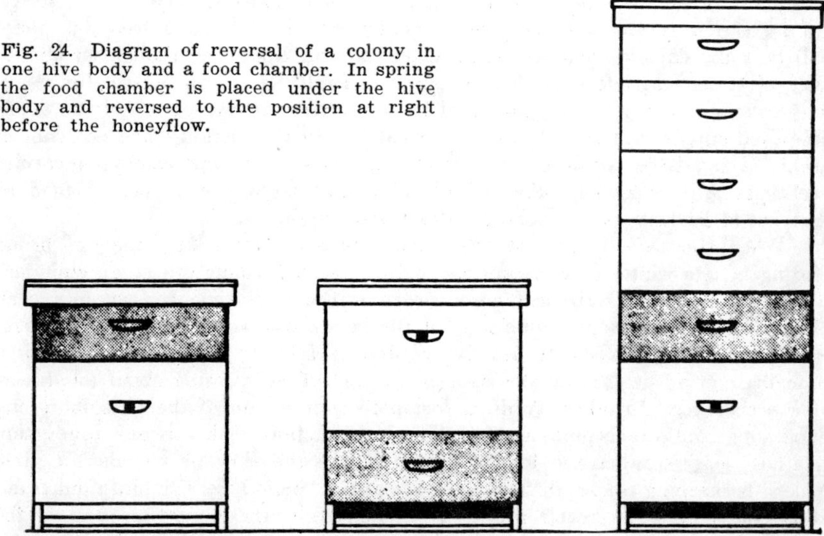

DEAD BEES CARRIED FROM HIVES

Q. On a warm day in early spring you can see many dead bees being carried out of the hive. What is the problem here?

A. You might often find a quantity of dead bees in front of a hive, especially after the bees have had a cleansing flight. These bees are the ones that had considerable age on them going into the winter cluster and they were not able to live through the winter. If they could not fly from the hive to die, they would drop to the bottom board and then be carried out of the hive by the younger bees. There is really nothing to worry about if your bees are healthy and if they had a relatively large force when they entered the cluster in the fall. If, in your opinion, the amount of dead bees appears excessive, keep a close watch on the hive for symptoms of disease or starvation.

SPRING BUILD-UP

Q. I have read that an overwintered two-story colony should have from 6 to 10 frames of brood at the beginning of the dandelion bloom. How many frames of brood should these colonies have had a month earlier, say as soon as the weather was warm enough to permit a first inspection? How much progress should they make in the month after the dandelion bloom?

A. A colony which would have 6-10 frames of brood at the beginning of the dandelion bloom would have had from 3 to 5 frames a month earlier. How far the colony will develop in the next month will depend upon many factors: queen, honey supply, pollen supply, location, weather conditions, etc.

Sometimes beekeepers rely too heavily on that first inspection of the colony. When they first open the colony in early spring or late winter, they might see frames of brood and some leftover honey and pollen stores and feel that the bees will succeed for the coming season. The period just before the main honeyflow is usually when colonies are most apt to exhaust their reserve stores. In the earlier season, relatively few colonies may have needed feed, but there will be some colonies that will require feed at this time even though an abundance of stores was left with them the previous fall. It is not unusual for populous colonies, with a large amount of brood to feed, to use up all of the available food supply and even to die of starvation. In this period when not enough nectar is available to supply the food requirements of the colony, a careful beekeeper will make sure that all colonies have an ample supply of food so that brood rearing can continue without interruption.

When the development of the colony starts with the beginning of brood rearing in late winter or early spring, with a vigorous young queen, an abundant supply of food, and sufficient comb area for brood rearing, the colony's peak of population may occur at the start of the honeyflow, and the ideal in management has been attained. It usually requires from 8 to 10 weeks between the time that spring begins and the start of the honeyflow to bring about this favorable occurrence. In other words, a fortunate conjunction of the peak in colony population and the beginning of the main honeyflow makes honey production relatively easy. Either too long a period before the flow or too short a time makes beekeeping more difficult. Too short a period for build-up may not allow the colony to reach proper strength before the honeyflow and if the period of build-up is too long, the colony may have reached its peak of population before the start of the flow, resulting in swarming conditions or loss of colony morale and resulting decline in honey-gathering ability.

BAIT COMB

Q. I have a hive of bees that doesn't want to fill the super above the brood nest. What can I do to lure them up into the super. They are building comb between the frames and on top of the frames.

A. Let's use a bait comb to lure them up into the super over the brood nest. This means taking a comb or two from a strong colony where the bees are building comb and storing honey. Place these combs near the center of the super you want the bees to occupy. The frames of foundation from the non-working colony can be put back in the colony from which the combs were taken. The bees of the non-working colony will then be more likely to be attracted up into the super where the two combs with honey in them have been placed. Generally, they should continue to work on the adjoining. frames containing foundation and should soon be working much better. We would also suggest that you check the nonworking colony for other problems such as shortage of food supplies or failing or dead queen.

SPRING FEEDING

Q. I am going to feed my bees and have two methods in mind: 1) feeding with a Boardman feeder and 2) pouring sugar syrup right on frames with drawn combs and then setting them in the hive. It seems to me that pouring the sugar syrup over the combs would save the bees a lot of time. Which method would you suggest?

A. We would suggest that the Boardman feeder or some other feeding method would be much preferred over pouring sugar syrup right on the frames of drawn comb. Pouring syrup right on the frames is not going to save the bees any time as they would have to take up the syrup, invert it themselves anyway, and then replace it back in the comb. Additionally, much of the syrup would be lost because it would run onto the bottom board, and there is also the danger of starting robbing activity.

Perhaps one of the best ways to feed your bees is to place sugar syrup in containers with holes in the lid and then turn the container upside down over the bee escape hole in the inner cover. An empty hive body or super protects the container from being spilled and the syrup is readily available to the bees whenever they need it. Two precautions should be noted in using this method, however. First you should make sure that the container is not placed completely over the bee escape hole so that a few bees can come out onto the inner cover and take up any syrup that might leak out of the container. Secondly, you should make sure that the cluster of bees is directly under the bee escape hole and if they are not, you should rearrange the combs so that the cluster is directly underneath the escape hole.

The plastic honey pails sold by bee supply houses make excellent feeders. You should use a small drill to make the holes in the lid of the pails so that the syrup will be accessible. If nails are used to punch holes, then sometimes the plastic will close up after the puncture and prevent the syrup from being taken.

Fig. 25. Two inverted 10-pound friction-top pails in position for feeding this strong colony.

POWDERY SUBSTANCE ON BOTTOM BOARD

Q. During the early spring I have often found a powdery substance on the bottom board. Does that indicate anything wrong in the colony?

A. This substance is the cappings from honey where the bees have opened the cells to use the honey, or it could be some wax or propolis they have gnawed on, trying to remove it from the hive. This should cause no problem but at the earliest possible time, it should be removed as this can only gather moisture entering the front of the hive and it can also be a likely spot where a wax moth could lay eggs. We have seen this wax almost an inch deep on a bottom board and later in the summer, there would be a crawling mass of wax moth larvae.

TWO EGGS IN ONE CELL

Q. During one of my spring colony examinations to determine if the queen was present and laying well, I found a cell that contained two eggs. What will happen to it? Is this unusual?

A. Finding two eggs in one cell is not unusual but does indicate that you are a pretty close observer of your bees. The bees will take care of the situation themselves. They will remove the extra egg before it hatches, but if not, will remove the extra larva after it hatches.

BEES DROWNING IN DIVISION BOARD FEEDER

Q. I have a division board feeder inside the hive with a long wood float in it but large numbers of bees seem to be drowning in the syrup. What should I do?

A. It is not normal for a large number of bees to drown in your syrup. Perhaps the float is too long and has caught on the ends of the feeder, causing the bees to have to go below the float to suck up the syrup. Sometimes it is helpful to put a piece of screenwire down from the top edge of the feeder to the bottom of the feeder. This way the bees have a sure foothold and can crawl out of the sugar solution more easily.

SPRING FEEDING

Q. I am feeding heavy syrup in division board feeders. Would it be beneficial to brood rearing if I would mix in a small amount of pollen supplement added to the syrup?

A. We would suggest that you feed the pollen in patty form rather than mixing it with the syrup in your division board feeder. This can be accomplished by mixing the pollen with a heavy two-to-one syrup mixture or with honey. Then, after the contents has formed into a firm patty it should be placed on wax paper in the hives.

Fig. 26. The division-board feeder at the front of this hive is convenient to use and holds ten pounds of sugar-syrup. Pollen is being fed in patty form on top of the frames. The pollen has been mixed with a heavy two-to-one syrup mixture, formed into a patty on wax paper, and then inverted over the frames. The bees will consume the patty and carry the bits of paper out of the hive. (Photo, G. W. Kloppe)

REMOVING LEFTOVER WINTER STORES

Q. In the spring after the bees have come through the winter, should you remove the super of honey you left on last fall for winter stores and replace it with an empty super, or should you leave the super on and let the bees refill it?

A. That super of insurance honey that you left on in the fall and wasn't used by the bees could just as well be left on the hive rather than remove it. Actually, more bees starve to death in the spring than in the winter. They appear to have enough honey source in the hive, but as the population of the colony increases rapidly their food requirements also increase. Oftentimes the spring flows are not sufficient to keep up with the demands of the colony and a colony that was well supplied on one visit to the outyard can be in trouble for food by the next visit. Leave that super on even though it has honey in it. An added reason for leaving the super of honey on the bees is that they are well prepared to care for it and if you tried to store it over the summer for use again as winter stores, you might have wax moth problems and additional storage problems.

WATER FOR BEES

Q. I have two colonies of bees in my backyard approximately 20 feet from my neighbor's dog kennel. My neighbor says my bees cluster around the water containers provided for his dogs and make a nuisance of themselves. What can I do about this?

A. The bees are using the kennel water supply as a source for their own needs and something should be done to maintain good relations. Honey bees use large quantities of water to dilute the brood food for brood-rearing purposes and to aid in cooling the hive by evaporation. This water is gathered by the bees and placed in convenient, temporary storage places in the combs, especially in burr combs and other comb constructions between the hive parts and along the top bars.

The need for water in substantial amounts is so great that the bees will obtain water from any outside source, frequently going to stock watering tanks, pumps, bird baths, lily pools, or similar places where they often become bothersome. This is particularly true in the spring and during periods when nectar is not plentiful in the fields. Bees may collect water from undersirable sources as well as from sources of clean water.

Some day it may be a regular practice that water should be given the bees in a feeder inside the hive where they may obtain it without having to fly in search of it. If your colonies are not located near natural sources of water, some sort of watering device becomes a necessity. It is best to provide water that does not stand in a container with a still surface but rather drips from a container onto a sloping board or trough where it is collected by the bees as it flows along. There is a strong suspicion that the spread of nosema disease is often aided by water sources which may become stagnant, such as low seepage places in the vicinity of an apiary.

In the spring, when bees first begin to get water, do all you can to prevent their getting a start in the wrong place, and to start them in the right place. Try a shallow container with cork chips spread over the surface so the bees will not drown as they take the water. Once you get the bees started with their own source of water in the spring, they should continue to collect water there.

SUMMER

SUMMER MANAGEMENT

Q. Once the Spring Management period has passed, what generally are the phases a model colony might go through during the remainder of the summer?

A. Model colonies are somewhat like model children — you may read and hear a good deal about them but they always seem to belong to someone else while your own (bees or children) are muddling along somewhere about average, here and there a success with a sprinkling of problems in between.

This book is about problems — no beekeeper has ever written in to ask why he was having so much success — so we'll outline a typical sequence of events occurring over late spring and the summer with suggestions to browse here and there in this book to see what problems other beekeepers have encountered and how those problems might be solved. The following material was taken from "Management for Honey Production," by G. H. Cale, and appears in *The Hive and the Honey Bee,* edited by Dadant & Sons. Words or phrases in bold type indicate areas of interest treated separately in this book.

When the development of the colony starts with the beginning of **brood** rearing in late winter or early spring, with a vigorous young **queen,** an abundant supply of **food,** and sufficient **comb** area for brood rearing, the colony's peak of population may occur at the start of the **honeyflow,** and the ideal in management has been attained. It usually requires 8 to 10 weeks between the time that spring begins and the start of the honeyflow to bring about this favorable occurrence. In other words, a fortunate conjunction of the peak in colony population and the beginning of the main honeyflow makes **honey** production relatively easy. Either too long a period before the flow or too short a time makes **beekeeping** more difficult.

If the honeyflow occurs in less than 8 to 10 weeks from the beginning of spring, there is often too little time for the colony to reach proper strength without the most expert care. It requires prolific young queens, an ample supply of food, and effective intelligent **management** — plus ideal conditions of weather, an abundance of **honey** and **pollen plants,** and freedom from **diseases** — to produce a maximum honey crop. If conditions are not ideal, it may be necessary to bolster colonies with queenless **packages** or to **unite** colonies prior to the honeyflow.

When the honeyflow occurs 12 weeks or more after the beginning of spring, the colony may have reached its peak of population before the start of the flow. It either will **swarm** before the main honeyflow or lose its colony morale and decline in its honey-gathering ability. To delay the peak of population in a long period before the honeyflow, colonies may be **divided** during the time of early flow from dandelion, fruit bloom, or other sources. A new queen is given to the queenless part and each division is allowed to grow until the honeyflow begins. The two parts may then be **united** or may be operated individually if there has been time for each part to achieve a satisfactory population of field bees for the flow.

This period before the honeyflow is also the time when many colonies follow a natural instinct to divide their populations through **swarming.** Because neither the parent colony nor the swarm will gather as much honey as the original colony, swarming should be prevented or controlled by proper management.

Assuming that your colony is entering the honeyflow with a strong population, has not swarmed, and is headed by a queen able to maintain the colony strength through the honeyflow period, there is little else to be done to improve the colony and it is ready for the honeyflow.

The remaining work for the summer is **supering** and maintaining the colony in good condition.

FALL

INFLUENCE OF QUEEN IN THE FALL

Q. I always assumed that the queen slowed or stopped egg-laying at the close of the honeyflow. Now I am told that an ideal queen is one whose egg-laying rate is high and continues well into late fall. Isn't this continued egg-laying and brood-rearing going to be using up valuable winter stores of honey and pollen before cold weather really sets in?

A. Producing brood in the fall when the honeyflow has ended does sound counter to good bekeeping practices until we remember that the colony needs young bees to go into the winter cluster. This is the time of year when the influence of the queen is greatly felt. Some queens continue their egg laying until late in fall, even after the beginning of cold weather. There is also a marked difference in the rate at which queens lay their eggs. Some queens reduce egg laying at a much quicker rate than others, although there may be some brood in the colonies throughout the entire fall period.

The ideal queen is one whose egg-laying rate is high and continues well into late fall, thus furnishing a populous cluster of young bees for winter. The stronger the colonies are with respect to a population of young bees before winter, the better they will be able to survive the winter period. A large colony of relatively young bees, well supplied with stores of honey and pollen, also will be able to rear brood in the latter part of winter and early spring. Such colonies will usually come through the following spring with little spring dwindling and, with proper management of their steadily increasing populations, can be brought to the honeyflow in optimum condition for gathering a maximum crop.

REQUEENING IN THE FALL

Q. I have been advised to requeen my hives in the fall. Wouldn't it be better to wait until the following spring so the queen would be fresh and new for the coming season?

A. Many beekeepers can benefit from fall requeening. The colonies have come through a spring and summer build-up and many queens will be old and worn out. A good practice is to tag the non-productive colonies during the flow. Such colonies should obviously be requeened. The proper timing is essential — and this time will vary from one location to another. As the fall season progresses, the brood rearing of the colony starts to decline. When brood rearing reaches a low level it is an ideal time to requeen. It is a matter of balance between the queen being removed from the colony and the new queen being introduced — neither is at the peak of egg laying and the worker bees are more apt to accept the new queen. The old queen is more easily found as egg laying slows and the colony cluster reduces in size. When you open the hive, start your examination for the old queen by removing one side comb and then splitting the cluster in the middle — working both directions. The old queen is killed and the new one introduced, usually using the mailing cage for introduction. As the colony is put back together, sprinkle a little sugar syrup over the top bars of the brood nest.

Colonies requeened in the fall of the year will pay big dividends for the beekeeper the following spring and summer. In the spring, such queens perform just as a new queen — building up large colonies capable of maximum honey production.

WINTER STORES

Q. With winter weather as severe as it is (Wisconsin), how do you make certain that colonies have enough winter stores?

A. This is always a difficult problem, especially as the weather becomes less predictable and seemingly more severe with each passing winter. You're not going to be able to guess the exact amount of winter stores needed, so if we have to be wrong, let's err on the side of giving too much rather than too little and losing the colony.

Start thinking in terms of preparing for winter during the fall honeyflow by not placing too many supers on the colony. Additional supers should be added only when the one on the colony is practically full of honey. This will prevent the bees from spreading the incoming nectar throughout more supers than they will fill. This becomes important when you remember that bees may not be able to leave the cluster in search of scattered cells of honey. There have been several reports of colonies starving even though there was honey to be had in the colony.

During the removal of the fall crop, one super completely filled with honey should be left on the colony. In addition to this super, the colony should have from twenty to thirty pounds of available sealed honey in the brood nest. This honey in the brood nest will usually be found along the tops of the brood combs. The honey in the super, plus the honey in the brood nest, should give the colony from sixty to seventy pounds of honey for their winter and early spring consumption.

If the fall flows are uncertain and little fall surplus honey has been gathered, the bees will have to be fed a sugar syrup for their use during the winter. The minimum amount of stores (either honey or sugar syrup) necessary for the winter consumption of the colony is fifty pounds. The beginner may find it convenient to take a pair of scales, similar to ice scales, and weigh the colony of bees. A complete hive with its bees should weigh about forty to fifty pounds with the top cover removed and any weight over that may be considered as stores. For example: if a colony weight is seventy-five pounds then it has thirty-five pounds of honey. To gain the additional fifteen pounds necessary for minimum stores, sugar syrup should be fed to the colony. The best feed for winter stores is made by mixing two parts sugar to one part warm water. A gallon of this mixture increases the weight of the colony about seven pounds so a colony weighing seventy-five pounds would need two gallons of sugar syrup to bring it up to the minimum stores for successful wintering.

Common practice is to leave a second hive body on each colony, one with sealed stores of honey and pollen being desirable. Record the amount of winter stores left on the colonies in your record-keeping system so you will have a guideline for the next winter.

Langstroth weights:
1 body — top — bottom = 40 lbs.
2nd body = 30 lbs.
1 super = 12 lbs.
Lid, telescoping = 8½ lbs.
Lid, flat tin = 6¾ lbs.
Lid, flat wood = 6 lbs.

BEES CLEANING CAPPINGS

Q. After extracting in the fall, is it safe to place the cappings 75 or 100 yards away from the beeyard so that the bees can clean them up? Will this cause robbing?

A. If the cappings are placed that far from the beeyard, we do not believe the bees would start robbing. The bees will clean up or rob the cappings and if hives are open in the beeyard while the robbing is going on, the robbing activity can extend to the hives. There is another thing you must be careful of and that is to be certain that all of the colonies were healthy and free of disease. Know disease when you see it and make sure that none of the colonies from which the honey was taken, were diseased. If even one colony has AFB and the honey was extracted with the rest, the cappings would spread the disease to all the bees in the yard.

PARTIALLY FILLED SECTIONS

Q. I guess I put on too many section supers. I did very well but I have quite a few partly filled sections. Some of these are partly capped. Would you please tell me how I can feed them back to the bees for winter feed?

A. The easiest way to feed comb honey back to the bees is to place the sections in a super under the brood chamber, or above the inner cover with the hole partially open so that the bees will take the honey out and store it in the brood nest. Many beekeepers who have only a few sections, uncap and extract the honey. Generally, section supers do not make good wintering space and I would not recommend their use over winter.

UNCAPPED HONEY AND BEES IN SUPERS

Q. In late fall I found I had supers above queen excluders that had a scattering of uncapped honey in them and bees that were difficult to remove. What do you suggest?

A. Go into the beeyard, remove the super and the queen excluder, put the inner cover on top of the brood nest, then put the supers on top of the inner cover with the hole open, recover the hive and leave it alone for a few days. If the temperature is around 35 to 40 degrees, the bees will have come down out of the supers to join the cluster. Then you can remove the supers without interference. One can also use such supers to feed weak colonies, or just set them out in the open if you are sure that all of your colonies are disease-free, and on warm days the bees will carry the honey back, storing it in their brood nests where they will use it for winter stores. Be sure that the open supers are about 100 yards or so away from the beeyard to prevent robbing.

FORMING THE WINTER CLUSTER

Q. When will the bees start to form their winter cluster and why won't the bees tolerate at least a few drones in the colony over the winter?

A. These are going to be approximate temperatures but they generally hold true in most areas. Understand that a windbreak or some other type of structure may be creating a mini-environment around the colony that could advance or delay the formation of the winter cluster.

At a temperature of around 57° F. the bees start to form their winter cluster. When the temperature has dropped as low as 43° F. all of the bees will have joined the winter cluster. With the formation of the cluster one of the first indications of approaching winter is the action of the colony in relation to its male population — the drones. The worker bees chase or drag the drones out of the hive where they soon die from starvation and cold. Since the drones have no useful task in a normal functioning colony they are a useless burden most of the time and would consume too much of the precious winter stores. The worker bees tolerate the drones during the spring and summer when there is a chance that they may be needed to mate with a queen, but the first hint of cold weather is the signal that the drones are no longer needed and the workers will chase them from the hive.

Fig. 27. A fine, fat drone (center) on the entrance to a hive in the fall. He will soon be excluded from the hive as food is too precious to waste on drones over-winter. (Photo, E. R. Jaycox)

Fig. 28. A windbreak for colonies is a good wintering aid.

Fig. 29. Close-up view of top entrance to a hive.

PACKING BEES FOR WINTER

Q. It gets rather cold here in the winter (Nebraska). Should my colonies have some sort of packing around them? What do commercial beekeepers do?

A. The majority of commercial beekeepers do not pack their colonies as the cost of the materials and labor for packing colonies would more than outweigh the loss of an occasional colony.

For many years the standard practice was to either winter bees in a cellar or to pack the colonies in a heavy coating of straw, leaves, or some other insulating material. The theory was that this packing would so insulate the colony that the natural colony warmth would be retained within the hive and allow the bees to move freely to their food reserve wherever it was stored in the hive. Some even insulated too heavily with the result that during warm days in the winter the colony was so warm that it was forced to move outside the hive. Generally, the trend has been to lighter packing or no packing at all.

Whether the colony is packed or not packed, two of the requisites for successful wintering are a reduced bottom entrance and a small top entrance. The entrance closer supplied with the original hive should be placed in the bottom entrance of the colony leaving only the smallest opening for the bees to fly from. It should not be necessary to close the entrance down so far until cold weather has set in for the winter. In addition to the reduced bottom entrance the bees should be supplied with a top entrance. A one-inch auger hole is bored just below the hand hold in the front of the super which is left on the colony. This top entrance allows the bees to fly freely from the top of the cluster during days when the temperature rises to 45° F. or more and also allows some circulation in the hive to help dissipate the excess moisture. If the bottom entrance should become choked with dead bees or other material the top entrance will then act as a safety exit from the hive.

WINTER PROTECTION FOR COLONIES

Q. I dislike packing colonies for the winter as it seems to be more trouble than it is worth. Wouldn't a windbreak work almost as well?

A. Although some beekeepers ignore wind protection, there are many who have begun to realize the importance of some kind of protection from the prevailing winter winds. This can be accomplished by several means: placing the hives on the southern slope of a hill, in a grove, or putting a snow fence or some other windbreak between the hives and the prevailing winter winds. Natural windbreaks are the best since they entail no extra management or work on the part of the beekeeper.

Winter protection is often one of the least worries of the colony, as colonies have wintered in extremely cold areas as long as the following essentials have been provided:

(1) A good queen early in the fall.

(2) Plenty of young bees.

(3) A minimum of fifty pounds of stores — honey, sugar syrup, or both.

(4) Reduced lower entrance and a 1-inch auger hole in super.

(5) Protection from prevailing winter winds.

(6) Exposure to winter sunshine.

FALL MANAGEMENT OF A CAPTURED SWARM

Q. Recently I acquired a swarm of bees in an unorthodox manner due to their location and a lack of time. The bees had build a large amount of comb about the size of a large grocery bag on the underside of a tree limb. Due to circumstances, I had no choice but to place an empty hive with bottom board and no frames under the swarm and saw the complete swarm from the tree limb and allow it to fall into the hive. Then I placed this in a permanent location and added a deep brood chamber with frames and foundation. After a few weeks, they are doing extremely well filling and capping the frames with new brood hatching daily in the upper and lower chambers.

My problem and questions — I live in Michigan with hard winters and it is now mid-October. Should I eliminate the mass of comb in the bottom chamber and replace it with another deep chamber with frames and foundation now or wait until spring? I believe the queen retreats to the bottom chamber when I start looking for her in the upper chamber.

A. We would suggest that you hold off on eliminating the massive comb in the bottom chamber and replacing it with new brood comb until spring. Doing this now just before winter comes would weaken the colony.

The queen is always hard to find, but if you separate the two brood chambers and place one over to the side, then you can look at one at a time without fearing that the queen will retreat into the other. In looking for the queen, you should look for the characteristic circle of attendant bees that form a court around the queen and that will give you a larger target to look for than just one queen bee.

Also a word of caution — Be careful not to crush or hurt the queen when you are searching for her. She will usually be located toward the center of the hive body where the brood is located.

CROWDING BEES TO FINISH COMBS

Q. When nearing the end of a season would it be advisable to crowd the bees into finishing combs by removing the center capped extracting combs? Do they start comb building in empty spaces late in a season?

A. Bees are not likely to start more comb building in empty spaces and on foundation late in the season. However, your method of removing the center capped extracting combs, and moving the partially worked combs toward the center of the super should help in finishing out the super. Chances are good, however, that it will simply remain unfinished. Frequently during the late season of the year, bees will ripen the honey in the cell and not even bother to cap it as the fall season progresses.

CLEANING UP WET SUPERS

Q. How would a colony receive the emptied or extracted frames of another neighboring colony at harvesting time? Would there be danger of robbing?

A. Any colony would be glad to have the wet supers from any other colony as it represents an easy way to store honey without the usual effort. Danger from robbing would come from honey on the outside of the supers or from placing the supers on a very weak colony that could not defend itself from invaders. Place the supers on toward evening to lessen the danger of robbing.

PLANNING THE NEXT SEASON

Q. It's too late to take any action this fall but I would like to know what in general I should be doing during the coming winter to get ready for next spring's honey season. I have ten colonies of bees and am thinking about expanding next year.

A. Provided you have already gotten your bees ready for winter and they have plenty of stores, your main job is done until spring. However, you should go ahead and get your equipment ordered and start putting it together so you will be ready next spring for the new packages that you order. Also, you might wish to devote some time to reading some of the new books out on beekeeping which probably can provide you with some interesting and helpful beekeeping pointers. At least, they will bring up some new subjects for you to think about. If you kept records during the honey season, the late fall and winter is a good time to go back over your records and notes and analyze your beekeeping operation. It might prevent you from making the same kinds of mistakes and might also open up new areas of beekeeping to you.

One of the main things, however, is to be prepared with equipment for the coming spring and avoid being rushed when the honey season does start.

COMB HONEY

FRAME SPACING FOR CUT COMB HONEY PRODUCTION

Q. What frame spacing do you recommend for ten-frame supers (6⅝") for cut comb honey production?

A. Standard spacing for frames is 1⅜" and this is the spacing that should be used while the bees are drawing foundation into combs. If wider spacing is used during this drawing of the combs, you're apt to end up with crooked combs and bridge and brace comb. After the combs have been drawn using the regular spacing, you're free to space them wider. In a ten-frame super for the production of cut comb honey you might space nine frames evenly across the super.

COMB HONEY

Q. I started beekeeping 2 years ago and the first year had one hive. I was interested in getting comb honey. I had 2 standard 10-frame hives with a box of empty combs of 28 sections. It took all summer but they filled 20 full sections. I was perfectly satisfied. In the early spring, bought more queens and bees and started them in 2 story 10-frame hives. Fed them good and everything was first class material. Not one comb did I get in the fall. So I tried putting the sections between the 10-frame so that the bees would have to go through the sections to get to the upper hive, but still they wouldn't make combs in the sections. Each hive had plenty of bees, so many that they were on the outside just crawling around. Why weren't those bees out getting nectar? I didn't get any comb honey the second fall either but the bees did gather enough honey to last them through the winter. What would happen if next spring I took the top hive along with 2 or 3 pounds of bees, gave them a new queen and on top of this put on a box of sections for combs? We have wonderful sweet clover here and I can sell a lot of comb honey if I can get it.

A. Comb honey is difficult to produce unless conditions are just about right. Not all bees want to produce comb honey, but most of them can be crowded into the supers with a fair degree of success. The queen appears to be important, so some selection is necessary if you want colonies which will produce comb honey. Colonies which, when crowded, do not go into the comb honey supers during a honeyflow should be requeened or used to produce extracted honey.

Our suggestion would be that you give the queen all the space she needs for brood rearing until the start of the main nectar flow. At that time, reduce the colony to one story which should contain the queen, brood of all ages, and a frame or two of honey. Shake most of the bees in front of this hive and add a comb honey super on top. When the bees start to work on the second super it should be placed directly on top of the brood nest and the first super placed above it. A third one may be added on top at this time. Remove the supers as soon as they are full and capped. The extra supers of brood should be stacked 3 to 4 high on a weak colony or used to make increase.

COMB HONEY

Q. I started working with bees in March of this year. My father, who had several hives when he was a boy, is helping me. We bought a shallow super which holds 28 split boxes. The boxes have the measurements of 4¼ x 4¼ x 1⅞ inches, and the wax is 17 x 4⅛ inches. We put it together, and put it on top of the hive with a queen excluder between the hive body and the super about the first of June.

A few weeks later we looked at the super and nothing had happened. They hadn't even gone up into it. So we put the super between the floorboard and the hive body with the queen excluder between the hive body and the super. A few weeks later we looked at the super and they hadn't built any comb in the super. We put the super on top of the hive and took off the queen excluder and poured some honey on the super and they didn't do anything. I would like to know how to get them to work on the super.

A. There are several ways to encourage bees to work in comb honey supers. We recommend that bees be allowed to become established in a bulk comb honey super first since the bees will start using it more readily. Then additional supers (sections) could be added between the well-worked bulk super and the body. Bait sections, sections which are partly drawn, may also help to draw the bees into the supers. Always put the section super above the brood nest, but never place it over a queen excluder. It is seldom that a queen will lay in a section super.

COMB HONEY

Q. You said in one answer that bees do not winter well in section comb honey supers. This worries me as I have two colonies with one section comb honey super on each one. What is the difference?

In the spring I suppose there will be brood in the supers. How will I get the brood out of them? Will they still be all right for section comb honey?

A. The problem with wintering in section comb honey supers is that there are too many divisions of wood for good clustering. If bees are to cluster well, they should be able to start on a few cells that are empty and should be able to move the cluster on comb in either direction at will. You may get away with it, but it is not the best practice. We would suggest you always have on hand at least a full shallow super for wintering each colony.

The best way to remove the bees and any brood that might be in the comb honey supers would be to place them under the hive body in the early spring, probably in early April. When the queen is again laying in the hive body, which is now on top, remove the comb honey super. It would be well to add a shallow super at this time as it will give the colony more room to grow. Large colonies are necessary for a good crop of honey, regardless of the kind.

These old sections will not be usable for a honey crop. Never use sections over except a couple of clean ones filled with comb which has not had brood reared in it. These may be placed in the super as bait sections to get the bees into the super more rapidly. These bait sections usually have a tough midrib and should not be sold.

COMBS

BURR COMB

Q. I am a beginner and have 5 hives of bees. I have them in two 10-frame deep supers for brood nests. When I manipulated them, I found quite a little burr comb on the frames. Should this be scraped off? If so, when should it be done?

A. Burr comb should be scraped off whenever you find it as it is only going to be a nuisance. The hive tool is excellent for removing burr comb and if you remove what burr comb you see everytime you examine the hive, your frames will be kept relatively free and much easier to manipulate. Save the wax. If you throw it on the ground, it may be the cause of robbing. A solar wax melter is an ideal unit for melting the bits and pieces of beeswax. It gets them out of your way and you'll be surprised at the amount of wax that will accumulate.

Fig. 30. Gleaming white newly-built burr comb built on the tops of the frames of a hive which had no inner cover. This colony should have been supered a week or more earlier.

FUMIGATION OF COMBS

Q. To save the combs, I put 8 frames and combs into a box and between each two boxes, a cardboard on which I placed both crystallized and powdered moth flakes. So far I have had good success, however, I have not been able to find out if these moth flakes have any effect on the combs. In the spring the bees do not seem to take to these combs too well. I do know they sure do ventilate the smell out that has gathered.

A. Moth flake (Paradichlorobenzene) will not damage honey combs although they should be well aired before being placed on the bees. If you have an air-tight room, the fumes will do an excellent job. The fumigated equipment should be well aired before you return it to the bees.

BRIDGE COMBS

Q. Would I get a lot of bridge combs if I changed my 6⅝" supers to 8 combs from 9? I uncap with a Fox Harrison uncapper and 9 seems a little thin, and I am thinking of using metal spacers.

A. In a good honeyflow, we do not believe that you would get a lot of bridge comb if you changed your 6⅝ supers, 10-frame size, to 8 combs from 9 combs. In a good flow, this would result in fatter combs which would be more suitable for uncapping with the Fox Harrison uncapper.

However, when drawing combs from foundation, except in a very good flow, you would get a lot of bridge combs constructed by such a spacing. When drawing combs, they should be spaced in a 10-frame super, preferably 10 to a super. But after the combs are drawn and in use, wider spacing would be satisfactory and we do not believe you would get a lot of bridge combs.

DRONE CELLS IN COMBS

Q. On recent examination of my hive I noted that two combs seemed to have an excessive amount of drone cells. Already noting that the hive contained many drones, I uncapped the sealed cells. In the next few days, the bees removed the drone larvae and cleaned out the cells. I do not suspect the queen of being sterile because there is a great number of worker larvae and sealed cells present, and also because I placed her in the hive only last fall.

Should I leave the combs containing the large amount of drone cells in the hive? Will the bees use the cells for honey storage? Will the queen continue to lay drone eggs in them? I fear an over-abundance of drones if she continues to do so. Would it be best to remove the combs and replace them with fresh foundation?

A. In response to your questions concerning the combs with large areas of drone cells, it is normal in the springtime for the queen to lay many eggs in such combs. Even though you have uncovered the drone cells so that the bees have pulled out the drone larvae and pupae, the queen will continue to lay in them until the cold weather of fall.

We would suggest that you do remove the drone combs and replace them with new frames and foundation. If there is no honeyflow when you do this, you should feed the colony sugar syrup so the bees can produce the new wax needed to make good combs.

One other precaution is to not put the frames with foundation between combs of brood in the brood nest. Rather, they should be placed between an outside comb of brood and the comb next to it which contains honey, pollen, or both.

If the drone combs are left in the colony, the drone population will become quite large. This will contribute to overcrowded conditions within the hive and this, in turn, could increase the possibilities of the colony's swarming.

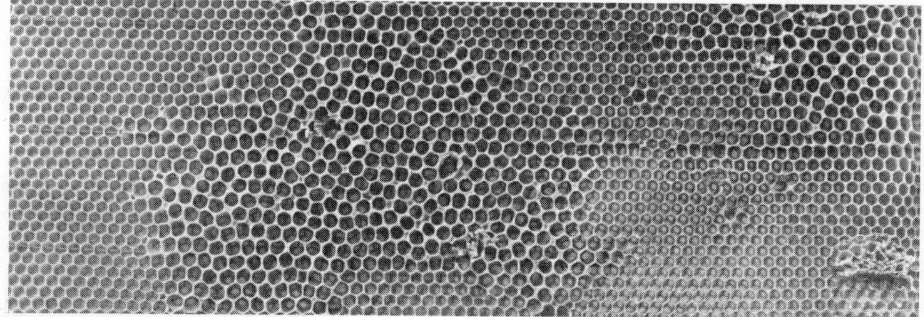

Fig. 31. The large amount of drone cells indicates this comb should be replaced.

REPLACING OLD COMBS

Q. When should old combs be replaced?

A. As long as the comb is straight and does not have too many drone cells in it, the comb should not be replaced, with one exception. This is if the comb had been used for several years and the cells had become small due to several generations of larvae having been reared in them. They may contain a variable number of cocoons and larval excrement, and the wax of the whole comb may be adulterated with pollen, cocoon fragments and other materials. Bees reared in small cells could be smaller than regular bees and may be less able to perform their duties as well as the larger bee. If a comb has a patch of drone cells as large as the palm of your hand, that is an excess and that comb should be replaced. Dark combs will not hinder the production of a colony, but each time a bee is hatched from a cell, that cell will become darker and smaller.

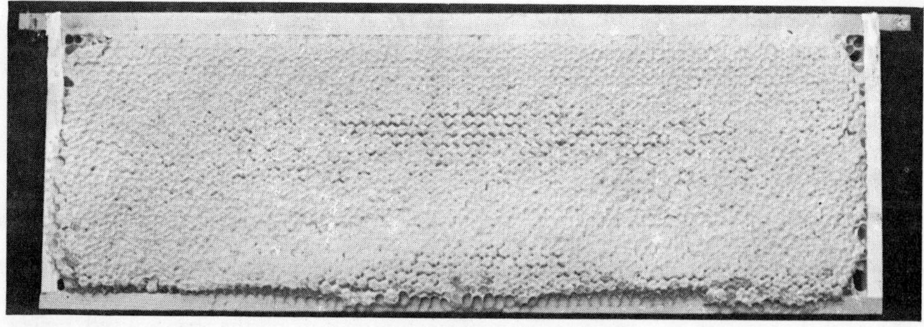

Fig. 32. While old combs may be used in the brood chamber, fine quality honey such as this deserves new combs.

BEES CLEANING COMBS

Q. I lost one colony of bees from starving. I cleaned up all the dead bees I could and burned them but there were quite a few bees left in the combs and a small amount of brood. Do you think if I gave the combs back to my other bees, that they would clean them up by taking the dead bees out or must I destroy the combs? The combs are nice and white and no mold. There is just a little brood on two of the frames and I wonder if the bees would clean that up too.

A. Bees are very efficient at cleaning out dead bees and brood from comb. However, dead brood should be checked by someone who can identify disease before the comb goes back into a colony. It is possible the dead brood may have been chilled, but it is also possible it was infected with foulbrood.

The Bee Research Division, USDA, Beltsville, Maryland, will be able to examine the brood and tell you whether or not it is infected. If the colony had disease, all the combs should be burned in a pit and the remains covered with one and a half to two feet of dirt and the hive body and other equipment should be disinfected by scorching or boiling in lye water.

COMBS WITH MOLD

Q. Two hives of my bees starved to death and I never expected it, as they had their hives full last fall and I never took any honey from them. Now I fed the bees some sugar mixed with water to give them a lift and found these two hives with dead bees. The foundation is drawn out, but it has mold on it. I wish to get some more bees to put in these hives and want to save this foundation so the new bees won't have so much work to do. Will the new bees take care of the mold and can I save it or what?

A. While it is not advisable, you can probably put a new package of bees on those combs that have some mold on them. A strong overwintered colony would clean up the mold on the combs quite quickly, however it may take your package bees slightly longer to do so. If the mold area is quite bad, the bees may chew the sidewalls of the cells down to the original midrib of the foundation and then rebuild the comb. Extremely bad areas may have the comb completely removed including the foundation base.

In your letter you state that two of your hives starved to death. This would indicate that possibly you do have additional hives. If so, we would recommend that you place these combs on top of your other colonies and let these overwintered colonies clean up the combs before using them for your package installation.

COMBS WITH CRYSTALLIZED HONEY

Q. If combs have crystallized honey in them will bees clean them out?

A. If you put a super on top of your bees with crystallized honey in them the bees will take out what they need to survive, then when a nectar flow starts in the spring, they will store new nectar on top of the crystallized honey. If you put the super on the bottom board and the hive body above it, the bees will carry all the crystallized honey up into the hive body and above the brood nest. They will not leave honey below the brood nest if conditions are so they can work it. This will take place in the spring, but during the winter months the bees will most likely leave the honey in the super. In fact, sometimes they go up into the upper hive body during winter, run out of stores and starve there. If feeding the crystallized honey back to the bees in a super, it should be uncapped and dipped into warm water before putting back on the hive. Just leave in the warm water long enough so that the crystals will absorb some of the moisture. This will help the bees in getting the crystals out of the cells.

COMBS DAMAGED BY WAX MOTH

Q. I have some combs that were slightly damaged by wax moth before I discovered that moth was in the hive. Can I use that comb again?

A. Yes, if the comb is not too badly damaged, the bees will draw out the comb and in many cases you cannot see where the moth damage was. If holes were eaten all the way through the comb by the moth, the bees will most likely put drone cells in that repaired part. However, you can use those combs at the outer edge of your hive. If they are too badly damaged, melt down and get what wax you can salvage.

COMBS FROM DEAD COLONIES

Q. Is it safe to use combs that bees have died in during the winter?

A. Yes, provided the colony was not diseased. If the colony that died was healthy but perished due to being queenless or running out of stores and left a cluster of dead bees in the hive, just shake off all the dead bees that you can, put another colony in the hive or catch a swarm and place in the hive. The new bees will immediately clean the combs and start a colony. However, if you find some cappings in the brood that are not fully opened, or what beekeepers call perforated cappings, and if you find a dried substance down in the cell, more than likely the colony was diseased and if you put another colony in that hive, they will soon die from the same disease and may even contaminate many other colonies. As a precaution, if you find a dead colony in the early spring, immediately remove it from your beeyard and enclose it in a building so that other bees cannot rob it out, at least until you can determine if the bees died of a disease.

SPACING OF COMB FOUNDATION

Q. What is the widest spacing that can be used between sheets of comb foundation before bees will build extra comb between them?

A. Dr. C. G. Butler of Rothamsted Experimental Station has this to say in his chapter in "The Hive and the Honey Bee:" Although they often appear to be spaced irregularly, the distance between worker brood combs is about 1⅜ inches, sometimes even more." H. C. Dadant in the same book has this to say: "At the time the combs are being drawn out from full sheets of comb foundation, the frames should not be spaced farther apart than 1½ inches from center to center. When fully drawn, it is common practice to space super combs as much as 1¾ inches center to center, resulting in thick combs of honey which facilitates uncapping them.

The spacing in the 10-frame hive is 1⅜ inches center to center. It appears that if sheets of comb foundation are spaced farther apart than 1½ inches center to center, that bees will build extra comb between them."

COMBS FOR WINTER STORES

Q. I have only one hive with two shallow supers producing cut comb honey. The supers do not have the comb sections in them but I have taken out four frames of honey. It appears as if the bees have given up on making honey as there are still empty frames in the second super. I want to make sure they have enough honey for themselves as I do not want to buy sugar for them. Should I just leave those supers on or should I take them off and replace them later on?

A. The fall flow in your area (Nebraska) is probably finished. Be sure to replace the four combs that you have removed. Leave the supers on for winter, unless the top one is nearly empty and if so, remove it. You should check carefully how much honey is stored in the brood combs. There should be a minimum of 50 pounds to see them through winter conditions in your area.

DARK COMBS

Q. I have been reading lately about the effects of dark comb on the honey that is stored in it. The studies have shown that dark comb, that which has had some brood raised in it, darkens the honey. This, in turn, lowers the value of the honey. Especially during a heavy brood raising period the queen frequently lays in empty areas of honey storage combs. My question is: Is there something I can do to these darkened combs to keep them from darkening the honey?

A. We do not know of anything which you can do to treat the darkened combs to prevent darkening of the honey.

It seems that the main things which can be done are of a preventive nature. For instance, you could remove all darkened combs from the honey storage supers, using them only in the brood nest or in supers which will have honey for wintering purposes.

The use of queen excluders early in the season would prevent the queen from coming up from the brood nest into the honey storage supers to lay eggs which will eventually leave cocoons in the cells, thus darkening the combs.

Another plan which you might try would be to keep a minimum of darkened combs in a super, say one or two, so that the amount of darkened honey will be comparatively small in relation to the total crop.

DISEASES

EFFECTS OF NOSEMA

Q. In what way does Nosema affect bees? What may be used to overcome it?

A. Bees with Nosema appear sluggish and crawl around on the grass and on the ground with swollen or distended abdomens. It usually appears in spring. At present the medicine recommended for Nosema is Fumidil-B which may be purchased with directions for use from Dadant & Sons, Inc., Hamilton, Il. In the case of Nosema, requeening is recommended and also setting the bees where they will have a good deal of sunshine.

NOSEMA DISEASE

Q. I have read that supersedure of queens often is due to nosema and that Fumidil-B is the correct medicant to use. Can Fumidil-B be fed to a queen in a shipping cage? Let's assume I buy one queen with about eight attendants. How would I measure the tiny amount of Fumidil-B, and how long could I keep the rest of the mixture in my refrigerator?

A. According to the work of Dr. Furgala of the University of Minnesota, one can feed a queen and her attendants with sugar syrup containing Fumidil. The suggested mixture is one level teaspoon per gallon of syrup. A queen and her attendants will consume about 1 drop from an eye dropper each day. You can keep this mixture in the refrigerator for a prolonged period without the Fumidil losing its drug activity.

Whether this would reduce or eliminate supersedure following introduction probably would depend on whether the queen was infected. If the queen was infected or was introduced into a colony with moderate infection and became infected, supersedure likely would occur. If the attendants were infected but the queen was not, the treatment probably would preclude infection of the queen but not of supersedure. The recommended way, when nosema is suspected, is preventive feeding right down the line — the cell builders, mating nucs and holding colonies in the South, and the new package colonies in the North.

NOSEMA DISEASE

Q. I am taking this opportunity to write you to request complete information on the effect of Nosema disease upon production of overwintered colonies and the results of feeding Fumidil-B. I am aware of the drug's value in package bee colonies, but I'm not presently sure of its value concerning established colonies.

A. Dr. Basil Furgala of the University of Minnesota is the one who has done most of the recent work on the effects of nosema disease and the beneficial results following feeding with fumidil.

The disease organism of Nosema apis has a buildup period in the fall after the weather starts to turn cold and bees become confined. The infected bees play host to a rather rapid multiplication of the disease spores within their stomach. Since the weather is cool and they are unable to fly out of the hive and defecate, the cleansing usually takes place within the colony itself. As the other bees attempt to clean or practice good housekeeping inside of the hive, they also become infected. Then, during the winter, the level of infection tapers back down. It is suggested that the heavily infected bees die and are not represented in the winter cluster. In the early spring as the bees increase their activities, the disease again builds to a higher peak. Again, it is normally tied in with the early spring conditions when cleansing flights are prevented by the weather. It is also aided by humid, damp conditions. Should this happen, the life of the adult bee is shortened, which means fewer bees to do the chores around the hive, both inside and out. As a result, spring buildup is usually delayed. In addition, it is not infrequent that the queens then become infected with the disease, and once the queen is infected, supersedure follows directly. Supersedure during the critical spring buildup period can cause a lapse in egg laying in the colony, which again leads to a slower buildup.

In addition, for those beekeepers who are reluctant to put the investment into the fumidil-B feeding program, many were urged to split the yards in half, using a fumidil feeding program on half of the yard and leaving the other half as a non-drug fed control.

At the end of two years, he has recommended that the beekeepers purchase some inexpensive used microscopes, and then do their own sampling to see if and when they need to reinitiate fumidil-B feeding programs.

AMERICAN FOULBROOD

Q. I am concerned about the use of drugs in preventing and clearing up an outbreak of AFB. I wonder if this is toying with the development of drug resistant spores, also the run down of a natural clean up ability of the bees.

A. Perhaps there is some possibility of the development of drug resistant strains of AFB. Even if this is a possibility, it is the lesser of several evils which might be AFB related.

If we may reminisce back to the days before antibiotics, it was not unusual to have from 10 to 20 percent of your outfit come down with AFB every year. The economic factors involved in this included the need for removing the diseased colonies from the regular producing yards, usually at night, to a hospital yard where they could be killed with cyanogas. The frames and combs were all burned, usually in a deep pit, and the remaining wooden parts of the hive were individually scorched with a blowtorch and scraped clean with a hive tool. In a 1500 colony outfit, this would represent about two men's labor for approximately a month to take care of the foulbrood equipment, plus replacing all the frames and foundation.

Now with the preventive feeding of terramycin, the incidence of AFB is very small. Since the action of antibiotics on the AFB bacillus is to prevent its growth or development, in effect there is no AFB. You are possibly acquainted with the fact that the common housefly did develop an immunity to DDT and this was probably due to mutations which permitted some small number of flies to live even though most of them were killed at a given time. Those flies which did live then propagated at a fast rate, thus bringing into being a race of flies which were more or less immune to DDT.

The same thing is very unlikely to happen with bees because the one queen in a hive is the only one capable of propagation. None of the thousands of

Fig. 33. American foulbrood in all stages, ranging from open discolored larva to cells with dried scales. The open cells contain sticky, decaying larva.

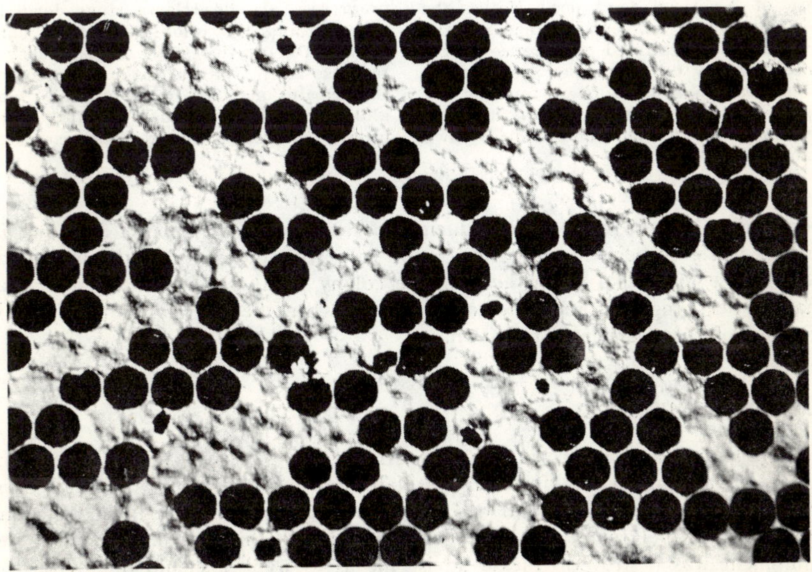

Fig. 34. The perforated and sunken cappings of American foulbrood. The bees have removed larva from the open cells but many of them still contain dried scales.

Fig. 35. Brood comb, with an advanced case of American foulbrood. Notice the perforated and sunken cappings on the cells and the scattered appearance of the brood. Many of the open cells, from which cappings have been removed by the bees, contain scales. When the larva has died with the foulbrood, it dries down into this scale.

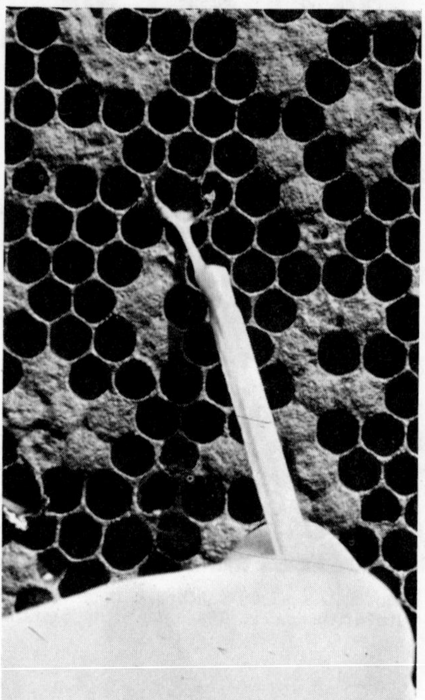

Fig. 36. As decomposition of the larva progresses, the dead larva sinks down in the cell and the color changes to dark brown. During this stage the larva has a ropiness to it. With a match or toothpick the remains may be drawn out like thick glue into fine threads. Typical dark brown foulbrood scales form on complete drying.

her workers are capable of reproducing so that should there be a mutation in any of these workers, the mutation would stop with the death of the individual bee. If the individual queen heading the colony is the one who has a mutation making her progeny susceptible to AFB, then when the colony is killed, that is the end of that particular mutation.

AMERICAN FOULBROOD

Q. I have a question in regard to AFB. I have never had any disease in my bees or equipment. However, last fall some of my supers that had been extracted were left outside and quite a number of bees were in them from another bee yard. I found out that there was AFB in this yard and some of the equipment was destroyed. Please advise me as to what I should do to prevent the spread of the disease to my yard especially the drawn comb that was in those supers.

A. It is doubtful that there is any danger of disease from the bees which robbed your equipment. It would be advisable, since there is AFB in the area, to feed your bees terramycin. Do this as early as possible in the spring and continue as long as the bees need feeding. Stop using drugs at least one month ahead of the expected honeyflow in order to prevent any drugs being stored in the supers.

If disease does occur it is possible to control it with terramycin. Terramycin is usually fed by mixing it with powdered sugar and dusting it down over the brood frames. Use one part TM25 with three parts powdered sugar. Four to six treatments about one week apart should do the job.

ELIMINATING FOULBROOD FROM WOODENWARE

Q. How much heat is necessary to eliminate foulbrood from hive bodies and frames? We're planning on placing the frames in a solar wax extractor and burning the inside of the hive bodies with a blowtorch.

A. Scorching the inside of hive bodies with a blowtorch is, of course, an old method for sterilizing the inside of hive bodies. However, placing the frames in a solar wax extractor will result in the beeswax being melted but we certainly would not consider the frames safe to reuse until they had been boiled in lye water. After such treatment, frames are seldom in very good shape for reuse and we would suggest burning the frames and getting new ones.

AMERICAN FOULBROOD

Q. During an illness I was unable to care for my bees and AFB got in and killed them. Is there any treatment that can be given to the hives and supers so as to render them for reuse? If I scorch them with a blow torch and wash them out with hot water containing lye would that make them reasonably safe for reuse? Should I feed the colonies a good supply of terramycin?

A. The treatment that you mentioned in your letter has been used in the past by beekeepers like yourselves who salvage the wooden goods from the hive. Combs are rendered out and the beeswax salvaged, then the frames and inner hive parts boiled up in lye water. It's pretty harsh treatment on the wood and also on the nails, but can work and give you salvagable material. The inside of the hive bodies can then be scraped and either scorched as you suggested with a blow torch or else burned out using something like kerosene in the center. The hazard here is that the fire may get away from you and scorch through too far into the hive body.

Within the last few years there has been work done on sterilization of the entire hive without damaging the equipment. The state of New Jersey seems to have taken the lead in this and has set up a sterilization chamber for this purpose. Ethylene oxide gas is the material used. Because it is potentially an explosive mixture, it is something at this time that would not be used as a home remedy. Perhaps you could contact Jack Matthenius, Jr., N. J. Supt. of Apiary Inspection, N. J. Dept. of Agriculture, Trenton, New Jersey, 08625. They could undoubtedly give you further details about the process.

Because of the general wide scale prevalence of spores of AFB, a preventive treatment utilizing terramycin would seem to be to your advantage.

AMERICAN FOULBROOD

Q. Would an escaped swarm be foulbrood free if it occupied a hollow tree year after year? My thought is that an infected colony eventually dies out.

A. You are quite right, an infected colony with American foulbrood would eventually die out. The chances are good that if a swarm or colony occupies a hollow tree for quite a few years, that this particular swarm or colony is disease free. American foulbrood is quite a killer and once a colony has the disease, it is only a matter of time until it completely dies out from the disease.

INFECTED HIVE ABSCONDING

Q. My lone hive was inspected and found diseased with foulbrood. The inspector told me it was curable and prescribed a drug medication. As instructed, I gave three treatments and about one week after the third treatment, all bees disappeared. Not even one dead or live honey bee or any other bee in or near the hive. Live brood was in the frames but no honey was in the entire hive. Why didn't I find at least one dead or live bee of any kind?

A. The most logical conclusion we could come to was that the bees became short on stores, especially honey, and after they had completely exhausted these stores, they absconded from the hive. Another possible explanation would be that the bees came in contact with some poisonous insecticide and died outside of the hive.

We do not think, however, that the drugs you used for treatment of the foulbrood disease would have injured the bees, especially in the small amounts that you used.

DOSAGE PER COLONY FOR AFB

Q. What is the prescribed mixture and the prescribed dosage per colony for the prevention of AFB? How often should it be administered?

A. The numbers of the two forms of terramycin — TM25 and TM10 — refer to the amount of the drug in each. For example, 2½ units of TM10 equals 1 unit of TM25. As terramycin quickly loses potency when put into a liquid, it is advisable to use it as a dust. For a small number of colonies, mix 1 part of TM25 with 3 parts powdered sugar. Four teaspoonfuls of the dust should be applied over the brood combs two or three times in early spring at seven to ten day intervals.

For large numbers of colonies, two pounds of TM25 is mixed with six pounds of powdered sugar. Four teaspoonfuls of this mixture should be sprinkled over the top bars of the brood combs at seven to ten day intervals.

CAUTION: Any use of drugs to prevent diseases should be confined to the period following the removal of the last honey crop in the fall until no later than one month before the first surplus honeyflow in the spring. No drugs should be given any colony during any part of the honey producing season. This is necessary so there is no possibility of getting any drugs into honey used for human consumption.

CHALK BROOD

Q. What causes chalk brood disease and what can I do to cure my colonies of it? I recently discovered it in several of my colonies.

A. The causative agent of chalk brood is the fungus *Ascosphaera apis*. Diseases of fungal origin are generally more prevalent in damp and cool conditions. For this reason, chalk brood is more commonly found in the spring and early summer. In some colonies, chalk brood can be found throughout the summer. Colonies rarely die from the disease, but in some cases honey yields may be reduced.

Susceptibility to chalk brood is at the maximum when larvae are about 4 days old. Remains of diseased larvae can be found in sealed or unsealed cells. Most affected larvae are found in the upright stage, however, in rare cases, larvae can be found in the coiled stage.

Chalk brood is transmitted via contaminated brood food. Once a colony is infected, the spores of the fungus can remain viable on the combs without causing any disease. When conditions for the spores to germinate become favorable, the disease could reappear. The fungus could also survive in the soil and somehow find its way into the food chain of the honey bees.

Chalk brood is seldom serious enough to require beekeeper treatment. No chemotherapeutic agent is registered for use against the disease in the U. S. In severe cases, the affected comb or section could be removed. However, in most cases, the diseased larvae are removed by the workers and the colony appears to recover spontaneously. Since cold and damp weather favor the development of the disease, colonies should be located to avoid moisture accumulation on the bottom boards or moved to dry locations. Sometimes enlarging the entrance of a colony aids ventilation and helps to dry out the colony, especially after a rainfall.

STABILITY AND STORAGE OF DRUGS

Q. How stable are the drugs used to control bee diseases? Also what is the best way to store them and how long can they be stored?

A. Terramycin® and Fumidil-B® are the two major drugs used to prevent bee diseases. Terramycin,® the trade name for Oxytetracycline from Pfizer & Co., Inc., is recommended for prevention of both AFB and EFB. Fumidil-B, developed by Abbott Laboratories, is very effective against nosema.

Terramycin has two forms — TM25 and TM10, the numbers referring to the amount of the drug in each. For example, 2½ units of TM10 equals one unit of TM25. Terramycin comes in amber bottles or in a dark container and the packages are dated for three years. As terramycin loses its potency in syrup rather quickly, the usual recommendation is to use it as a dust in a mixture with confectioner's sugar.

Fumidil-B is very effective against nosema and is fed to the bees in sugar syrup. A small package (0.5 grams of activity) is used in five to six gallons of syrup to get the desired concentration. The larger package (9.5 grams of activity) is used in from 100 to 120 gallons of sugar syrup.

Fumidil-B is not as stable as other drugs and the packages are dated for a period up to two years. Fumidil-B should be stored away from light. In fact, most chemicals store best in cool, dark, dry places.

Dr. Furgala has shown that Fumidil-B fed to wintering colonies in sufficient volume of syrup suppresses infection through the winter and that syrup taken from treated colonies in the spring still has antinosema activity.

Fig. 37. Dusting for disease control. Inspection sticks will be burned in the smoker and sanitary measures will prevent disease from being spread to other hives.

DISEASE CONTROL

Q. Please explain the use of terramycin for disease control?

A. Terramycin has been used not only for European foulbrood, but it has also proved to be a good preventative for American foulbrood. Terramycin is a very effective control for the prevention of American foulbrood when fed to wintered colonies in late summer to mid-fall and early spring to late spring. Terramycin can also be fed to package colonies from mid-spring to late spring. As terramycin quickly loses its potency when used as a liquid, it is advisable to use it as a dust. It is good for a few days only and must get into brood areas at once.

In using terramycin in the control of both American foulbrood and European foulbrood, it should be fed as early as possible in the spring when brood rearing is well underway, at the rate of one level teaspoonful of terramycin either dusted on the combs or in at the entrance and up among the combs or in the feed. A typical formula is TM25 mixed 1:3 with powdered sugar.

The treatment should be repeated in about three weeks. The terramycin mixture will cost about 5 or 6 cents each time you use it, but, of course, it has the advantage of being a preventative for both American and European foulbrood.

QUANTITIES OF TERRAMYCIN DUST

Q. How much of the formula of Terramycin dust is needed per colony of bees as a treatment to prevent the occurrence of AFB? I have read the proportions many times for the mixture, but never the quantity needed per colony.

A. It has never been determined just how much terramycin is needed per colony. There are too many variables such as condition of the queen, strength of the colony, the amount of the disease, etc.

We find that three to four treatments during late summer to mid-fall and another three to four treatments during early spring to late spring are usually sufficient to free a colony of all visible signs of disease.

DISEASE OF THE LARVAE

Q. My package colony has something that seems to be attacking the larvae. Much of the brood is dead before the cells are capped. The color is reddish yellow cast. I tried dusting them with terramycin with no results. Can you tell me what is the trouble?

A. Without knowing more specific details about your trouble, it is difficult for us to tell you what might be wrong. We suspect, however, that it is European foulbrood but this should have responded to the drug Terramycin.

We do have a suggestion. Send a 2-inch square piece of the comb containing the diseased larvae to Dr. H. Shimanuki, Microbiologist, Bioenvironmental Bee Laboratory, USDA, ARS, Beltsville, Maryland. This laboratory will make an examination and will report their results to you at no cost.

EUROPEAN FOULBROOD

Q. Is European foulbrood carried over from year to year in the combs or honey after once cleaned up?

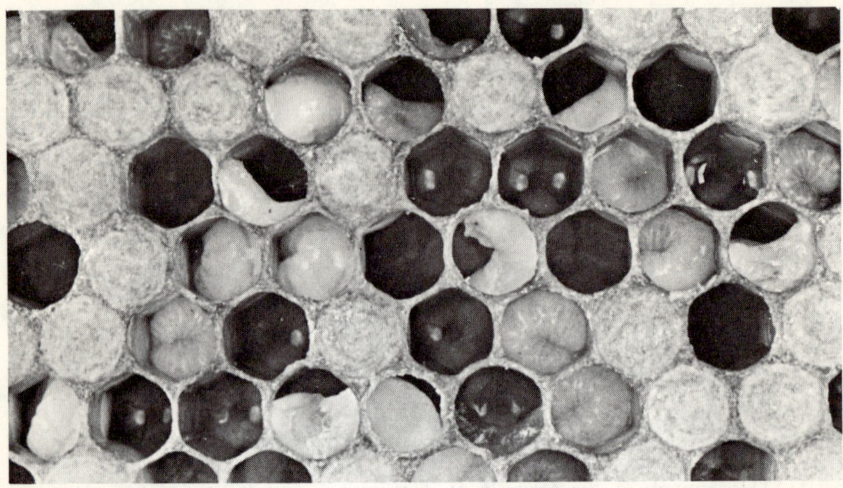

Fig. 38. Closeup view of European foulbrood damage. Some of the larva appear almost normal, others are infected and discolored, and some are dead masses in the bottom of the cell.

A. According to our information, EFB sometimes is called a "stress disease," appearing in areas with unfavorable weather or crop conditions. *Streptococcus pluton,* swallowed with food, grow in the mid-gut of the bee and are evacuated with the feces. The *Streptococcus pluton* is deposited mostly at the base and cappings of cells. Sometimes under dry culture, it will remain alive up to three years. Most of it, however, is removed by housecleaning. Some of the disease may find its way to the colony through food given to larvae and may be able to repeat its cycle.

FEEDING HONEY

Q. Is it all right to feed honey that I melted up, back to the bees in the spring?

A. Too many times articles appear in the bee journals, saying to feed honey back to the bees, with never a mention that the honey that is fed back to the bees may contain the germs of bee disease. It is doing the beginner beekeeper an injustice by not explaining this. A lot of beekeepers, especially those who have not had bees long, do not know bee disease and wouldn't know how to look for it. By feeding contaminated honey the disease is spread from possibly one colony to the entire yard. When we say contaminated honey, we do not mean that the honey is injurious to the human race. Statistics show through laboratory tests, disease germs that affect the human being cannot live in honey over a period of but a few hours. But germs injurious to the bees can and do live on in honey. If a colony dies out, and you do not know the reason it died, do not feed the honey to other colonies, or allow your bees to rob out the dead colony. Be sure that you know bee disease when you see it. As an added precaution, if surplus honey is fed back, add a sufficient amount of antibiotics to the mixture.

DIVISION OF COLONIES

DIVISION OF COLONIES

Q. First, we received a badly damaged 2-pound package of bees late in the spring. We nursed them through the winter and they did not make any move to swarm. In the spring we purchased two new queens and tried to divide the hive, giving each new colony a new queen. One new colony lived but did not seem to increase in number as they should have. We were getting ready to order another package when the other colony swarmed. We made an effort to gather in all of the swarm that we could. They have taken more food than before and we hope they will live. Now should we buy another new queen or queens or what?

A. If both of the colonies are doing all right, leave them alone but watch them carefully for signs of shortage of food, disease, or failing queen, etc. You should have fairly new young queens and they should be prolific layers which would ensure a good build-up into strong colonies, provided there is no disease and the food supply is adequate.

Generally, we prefer to requeen as needed rather than replacing queens in the hopes of solving problems that may be due to some other situation. Keep a close watch on the pattern of brood as a clue to the need for requeening. A good queen will be laying solid patterns of worker brood while failing queens will often lay in scattered patterns or excessive amounts of drone brood.

Fig. 39. A comb of bees and brood that could be used in dividing colonies to make increase. Make the divisions in early May about 6 weeks ahead of the honeyflow during fruit bloom to give the new colony a source of food. The colony will also need to be fed until they are strong enough to provide for themselves.

DIVIDING COLONIES TO MAKE INCREASE

Q. I am getting five new queens this spring in May. I have five good queens now but can I take the old queens and two or three frames of brood and some honey and put them in another new hive and make another colony of bees out of them so they will make enough food for next winter?

A. The procedure you describe of taking two or three frames of brood and honey for starting a new colony is rather common practice. The success you have in doing this will depend on how strong the colonies are this early in the season, so they can afford sparing those three frames of brood without imparing their buildup for the honeyflow.

A lot of this brood making is done in the South where their season is earlier, and they can then move the division or nuc to the North for buildup in time for the flow. Where it is done in the North, usually a program of stimulative feeding is started fairly early in the spring or even during late winter to make certain that ample brood is there.

You will find that the ability of the nuclei to develop into a full colony is more satisfactory than with a package of bees. The emerging brood from those combs helps carry the bees over the critical dwindling period and gets the colony going faster.

DIVIDING COLONIES TO MAKE INCREASE

Q. I would like some information on how to divide some hives of bees. I would like to divide them and use hybrid queens in the new hives.

A. Take most of the sealed brood with adhering bees and a new queen and place them in a hive in a new location. The new half, having mostly young bees, will be less likely to kill the new queen than would the old bees in the original half. Many of the field bees will return to the parent hive and will care for the young brood left and will not harm their original queen.

Moving the bees about two miles will eliminate most of the drifting between the two colonies. Make most of the divisions in early May about 6 weeks ahead of the honeyflow; this is also fruit bloom, giving the bees a source of food. The divisions should be fed until they are strong enough to provide for themselves.

DIVISIONS USING A QUEEN EXCLUDER

Q. I have five colonies and would like to make a new one. Can I take a queen excluder and put a fourth inch rim on top and put half of the bees above this? How long must I leave it on with the new queen above? I have two story hives.

A. Yes, you can take a queen excluder and put a rim on top of it when you make your division. We assume that you plan this to be the entrance for the upper colony. It probably won't be easy to requeen the upper unit, since all the bees will be of one colony, passing through the excluder at will. We personally would prefer a double screen so the bees are less likely to get together. You can remove this upper colony at any time after it has become organized. It must be removed some distance (over 1 mile) or all the field bees will return to the old stand. A second choice would be to remove the bottom hive a short distance and place the new one on its stand. In this way you will build up the new colony very quickly as the new colony will pick up the returning field bees from the original hive.

DRONES

ANCESTRY OF DRONES

Q. My grandfather used to really get me confused with the saying, "A drone never has a father, but always a grandfather." Would you explain this?

A. The saying really is rather confusing until you know about the ancestry of drones. All female bees, whether queens or workers, have fathers because females are produced from fertilized eggs. The birth of a drone can come about in three different ways but remember that in each of the three cases the drone is going to be fatherless because drones develop only from unfertilized eggs. In each of the three cases, however, the drone will have a maternal grandfather because the unfertilized egg that produced the drone was laid by a female and all female have fathers.

1) The queen bee, after mating with a drone, can lay two types of eggs, fertilized which produce workers or queens, and unfertilized eggs which produce only drones. The unfertilized egg received no sperm and hence, no father for the drone in this case of parthenogenesis, or virgin birth.

2) A virgin queen, one who has never mated with a drone, can lay only unfertilized eggs which, again, produce only fatherless drones.

3) In certain conditions such as a colony becoming hopelessly queenless or in other abnormal situations, a worker may lay eggs which will produce drones since the worker had never mated and the eggs were unfertile.

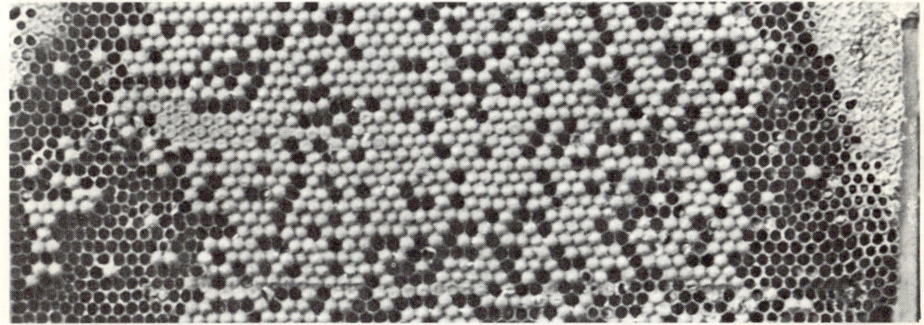

Fig. 40 Drone comb. Once combs have been "drawn" into drone cells, the bees cannot change the cells back to worker size. A sure cure is to melt up the comb and replace it with new foundation.

DRONE CELLS IN COMBS

Q. What action do you recommend for beekeepers who find several frames of solid drone cells and no worker brood? I recently put the bees to death in two such hives, for fear of flooding the beeyard with honey consuming drones.

A. The only way you could get rid of those drone cells would be to melt up that comb and insert new foundation. Once bees do change the cells from worker to drone, they cannot change back. The bees in the colony could be united with another colony or killed. If the colony consists of nearly all drones, it would probably be just as well to kill them as the workers would be old.

LACK OF DRONES AFFECTING YIELD

Q. I hived a 3 lb. package of bees in August. There were a few drones at that time but I have not seen any for the past three weeks. The hive is full of bees and from the looks of the brood coming off, the queen must be working overtime. I have noticed a few drones cells but none have come off as yet. Could the shortage of drones be due to the shortage of nectar? Should I encourage the rearing of more drones to prevent my yield being affected?

This is my first hive and I have had no experience in keeping bees but am learning by the trial and error method. I believe it to be one of the most fascinating hobbies one could have. If more people would try it there would be less sales of tranquilizing pills.

A. You should consider yourself fortunate to have so few drones. Although there is a theory that colonies permitted to rear considerable numbers of drones will be more content and thus, have a higher yield than colonies whose drone-rearing activities have been restricted, the theory has generally given way to the practice of restricting drone-rearing whenever possible. The natural instinct of the bees will prod them into rearing some drones prior to the swarming season and even if your colony contained perfect worker brood comb, which is very rare, the bees would tear down some of the worker cells and rear some drones just to satisfy their instinct. The rearing of any brood is expensive from the standpoint of consumption of honey and most beekeepers would prefer as large a number of workers as possible. Colonies that do rear an excessive number of drones should be watched closely as the queen may be failing, which would mean that her eggs are unfertile and will hatch into drones rather than workers.

A package hived in August would be in the period when colonies normally slack off on drone production. In fact, this, on through to fall, is the time of year when it is difficult to get drone production in breeding programs.

DRONES DYING WITHOUT CARE

Q. Is there any explainable reason why drones die within an hour after being placed in a vented jar?

A. The drones apparently need continued care and feeding from the worker bees. Although drones will occasionally feed themselves from an open nectar cell, they solicit worker bees quite actively for care and tending. We have noticed the same thing ourselves here in our breeding program and that about an hour after the drones are removed from the hive, they start becoming extremely feeble. At that time, we place them back into the hive in a caged chamber, and let the workers revive the drones for us.

DRONES IN THE WINTER CLUSTER

Q. Why do some colonies carry drones through the winter while others do not? Also is a drone reared from a laying worker egg capable of mating?

A. Colonies carry drones through the winter only when they are queenless or have virgin queens. They would keep the drones for mating with the queen. Otherwise, they would drive the drones out of the entrance before winter to die from starvation or exposure. Yes, a drone reared from a laying worker egg would be capable of mating since he is a normal male.

ENEMIES

BLACK ANTS

Q. I have caught a swarm of bees and put them in a hive. And since then small black ants have taken over my hive. I lifted the hive up and put burned motor oil in a pan and then I put some bricks inside the pan of oil and set the hive back in place on the bricks, but I still find the ants crawling on the outside and on top of the gum. I have searched through the gum and I can't find any of them inside it. I have killed all I could find on the outside today, but tomorrow I can find some more. Do you have any suggestions about how I can get rid of the ants?

A. There really isn't anything that anyone can do to kill ants that won't also kill bees. Probably one of the best suggestions is simply to remove the hive temporarily and sprinkle some kerosene around on the ground. Then put the hive back in position. Kerosene acts as a kind of a repellent for the ants and they tend to stay away at least until the effects have worn off. The presence of ants around and on the outside of a beehive is no cause for any alarm. As long as a beekeeper keeps the colony strong and healthy, the ants will not actually invade the hive or cause any damage.

BARN SWALLOWS

Q. Having a few bee colonies since last year near an empty barn, I find the bees don't do well at all, and I am wondering if the barn swallows, martin swallows and the king birds eat the bees?

A. Although the birds that you mention may pick off an occasional bee, we understand that they normally do not use the bees as a steady diet. The complaints in this regard from other readers have been lacking, so at this point we'd be suspicious of this as a cause.

BEARS

Q. How does one bear-proof an apiary (Oregon)?

A. It's a serious problem in bear country and we have seen more gadgets and ingenious solutions than there are bears. One such solution called for a small transistor radio tuned to a radio station catering to loud rock music. The theory is that the bears will assume people are present when they hear the music and will not stay around for a possible encounter.

An electric fence seems to be one of the most practical ways to bear-proof an apiary. We have used five strands of wire all carrying the electric charge. Net wire of some type lying on the ground outside the fence helps to ensure a ground. The wires should be about one foot apart in the fence.

An outyard, of course, would require a hot shot battery. Yards near buildings could use 110 volts as the source of power to be run through a fence charger but warning signs would also have to be posted to prevent harm to humans and a possible lawsuit. The U.S.D.A. has done some work on an electric fence for bears.

Fig. 41. An attempt to bear-proof an outyard. Note the wire laid on the ground before the electric fence to ensure solid contact.

Fig. 42. This thief was determined and even the flash of the camera did not deter him from attacking the hive.

EXTERMINATING YELLOW JACKETS

Q. Is there any way of exterminating yellow jackets without injury to my bees?

A. Probably the most common yellow jackets build their nests in the ground and can be exterminated by placing some insecticide in the hole to their nest. In fact, the place to attack and destroy yellow jackets is at the nest and away from the hive. There are insect killer products on the markets — check your beekeeping supply house. We should also remind you that insecticides are toxic materials and should be used at all times as recommended on the label. In case you do have to get at the yellow jackets in the presence of bees, try to make the application on a rainy day or at night when the bees are not flying.

DRAGON FLIES

Q. At one of my beeyards this summer I noticed more than once quite a number of dragon flies in the air. I would like to know if dragon flies kill a significant number of bees, and if so, is there anything to do other than find a different location. I got a fair yield of sumac honey from some of my hives there, so I would not want to move except for some good reasons.

A. Although dragon flies are voracious predators of other insects, we had not heard of them being a problem as far as worker bees were concerned. We have heard reports, particularly in the southern areas, where they do take quite a toll of virgin queens on their mating flights. However, we hadn't heard where they had decimated the field population of colonies. One would think that the numbers of dragon flies would have to be extremely high to have any effect. The presence of large numbers of dragon flies in your bee yards may have been due to the presence of the honey and the numbers of insects other than bees that might have been drawn to the honey. The dragon flies may have found your honey-laden hives to be a convenient lure for unsuspecting insects.

Fig. 43. Skunks are insect eaters in nature and they cause considerable disturbance in colonies during their nighttime visits, scratching at the entrance and eating the bees that respond to the scratching.

PURPLE MARTINS

Q. We do not yet have bees, but would like to. We have so many questions. One, would our martin colony bother bees?

A. We would certainly like to encourage you to go ahead and get at least one or two hives next year. At this stage, you cannot help having many questions, but that is the beauty of beekeeping as a hobby — the answers to your many questions will come from actually looking at and working with a colony of bees.

As with most new things, you will probably make mistakes, but that is part of the fun of the hobby and an added attraction to beekeeping as a hobby is the fact that under normal conditions, you can expect to receive a harvest of honey, which is certainly worthwhile in itself. In addition, the benefit of the bees in pollinating various food-producing plants like fruit trees is worth far more than the honey crop which the bees will gather for you.

We doubt very much that the martin colony would have any effect upon the colony of bees. Honey bees have survived under all sorts of conditions for thousands of years. Probably the martins would be more apt to catch drones than they would the worker bees. The drones contribute nothing to the welfare of the colony except for being available for mating should the colony raise a new queen. They neither gather honey nor pollen. In fact, all they do is eat pollen and honey which the worker bees have brought in. If the martins ate a few of the drones it would be a help to the colony. A beekeeping friend told one time about watching the bluejays teaching their young ones how to catch the drones in his beeyard. The bluejays only showed up with the young ones during the period shortly after dinner, like from 1 on to 3 o'clock, when the drones are flying best. He said it was quite obvious that the old bluejays were very careful in catching only the drones. Perhaps the jays preferred the drones because they do not contain bee venom which the worker bees contain.

WAX MOTHS

Q. How do you keep wax moths out of hives and is there a way to save the bees?

A. At the risk of evading the question for awhile, the best prevention against the wax moth is a strong colony and apparently your bees did not have time to build up or something happened to them to cause them to become weak. Consequently, the wax moth entered and laid eggs which developed into wax worms which do the real damage. In such a case it usually is not long before the wax worms have complete control of the situation and everything is ruined.

We would suggest that you order package bees earlier to arrive about the time of the first fruit bloom. This gives the package colony a chance to build up to strength in the cooler weather of spring before the moths become too prevalent. Keep the entrance reduced until the colony has become large and populous and normally you will not have any trouble from moth damage. Once the wax worms are established, there is nothing you can fumigate the combs with that won't also damage the bees. If you can shake the bees from the combs, letting them drift to nearby colonies, any combs worth saving can be fumigated with paradichlorobenzene, available from Dadant & Sons, Inc. Then in the spring, air the hive bodies thoroughly before returning them to the bees.

Fig. 44. A worker escorts the corpse of a wax moth larva to the door. The best possible defense against the wax moth is a strong colony. (Photo, Bee-Wolf)

EQUIPMENT

BEE TRAP

Q. Please send me information on making a bee trap for catching wild bees.

A. Bee hunting apparently is great sport. Doubt that it pays very well but then what hunting does?

There are no printed plans for a bee trap. Everyone has his own ideas as to what one should be. Traditionally a trap consists of a small box containing a piece of comb which is filled with honey syrup flavored with anise for scent. This box has a hole in it covered by a sliding tin which leads to another box with a sliding glass top. Catch a bee gathering nectar (not pollen) and place her in the bottom box. Close the tin. After she has filled herself with honey allow her to come into the upper box and escape. Watch her flight. She should return with other bees in a few minutes. If they follow the same line it is possible, by controlling the number of escaping bees to establish a line. Move the trap from time to time along this line until the tree is located.

If a thousand bee hunters read this letter, we could find a thousand ways to do the job. Good luck, maybe we both can learn more about this sport.

BOTTOM BOARDS

Q. The bottom board of my hive has two sides to set the hive upon. One side raises the hive up higher from the board than the other side does. If this is true do I have to turn all the boards over until the hive is resting on the shallow side?

A. We use the deep side of the bottom board the year around. We do use an entrance cleat in the opening for the winter. You may, if you wish, reverse the bottom board for a winter entrance.

Fig. 45. The bottom board serves as the floor of the hive. It may be reversed for a more shallow clearance. The entrance block provides for different openings.

BRANDING BEEKEEPING EQUIPMENT

Q. Recently I have heard a great deal of conversation about branding beekeeping equipment. Why brand equipment and what units are available for branding?

A. Many people have had their colonies "Bee Rustled" in the past few years. Now that honey is at a profitable level, rustling is becoming more common. Most people agree that branding of bodies, supers and frames helps deter theft of equipment because it can easily be identified. The problem is so severe in some areas that non-branded frames have been stolen leaving only empty branded supers and bodies.

Brands are available from the economy brands with three ⅜" letters to a propane brand with up to five 1" letters. The small economy brands must be placed in a heat source (open fire, burner, gas flame) to brand. The propane units run off tanks or disposable bottles. In addition to these units an electric model is also available. It has up to five 1" letter capacity and is convenient since it can be used wherever there is an electric outlet. It is, however, less efficient and takes longer to build up heat between brandings.

SHALLOW SUPERS FOR BROOD CHAMBER

Q. Have you done any experimental work using just the shallow supers for the brood chamber? If so, how did they work out?

A. We have used shallow supers for the brood chamber in the Dadant apiaries. This was a method long practiced and advocated by the late Dr. C. L. Farrar, former chief of the Apicultural Research Branch, U.S.D.A. and former head of the apiculture research work at Madison, Wisconsin. This work has been continued by Dr. Floyd E. Moeller. Both speak highly of the system and point to the advantage of having all interchangeable equipment. It is a system based on manipulation of individual shallow bodies, rather than the manipulation of individual frames or combs. In addition, the shallow bodies are easier to handle weightwise.

We use this size of frame in the queen nuclei used in our bee breeding program. We do not prefer it for the brood nest because there is so much interruption of comb area by the top and bottom bars and the space in between. We like to have a more compact brood nest and use the standard 10-frame bodies; most of this is a matter of personal preferance. We have heard from one woman who is an excellent beekeeper who uses shallow supers exclusively because of the weight factor and her colonies were producing very well.

BUILDING EQUIPMENT

Q. I plan to start on a small scale in spring more or less as a hobby. How many colonies would you recommend I start with? I am familiar with wood and metal working machines and I plan to build my own equipment. Where can I get the necessary drawings for the complete hive, extractor, etc.

A. We would suggest that you start your beekeeping with two colonies of bees and that you purchase two-pound packages to be installed in early April. You can go on from there depending upon how you enjoy this hobby, but we're certain you will find it fascinating.

If you really wish to build your own equipment, buy or borrow a complete outfit and make exact copies. The problem with beekeeping equipment is that the dimensions must be accurate or you will have trouble as long as you use the equipment. If the space is too large, the bees fill it with comb; if the space is too small, they glue everything tight with propolis. We seldom recommend that beekeepers build their own equipment since the savings is not great and the chance for error is fairly high. In beekeeping, you will find your time better used if you buy the equipment and spend your time working with the bees.

REDWOOD AS MATERIAL FOR HIVES

Q. During the past few months, I have gathered a lot of scraps of redwood, from which I made a number of bottom boards for my beehives. Since then I have been advised by a friend who works for a redwood distributor that redwood is bad for apiary supplies. He based his opinion on the fact that no insects live in the redwood forests and that he had also been told what he told me. What is your opinion? Also would it help if I painted the inside of the bottom boards as well as the outside?

A. We are not an authority on redwood nor what qualities it has that might keep insects out of redwood forests but we suspect that your friend is either misinformed or is overstating the case. We know that redwood has been used fairly extensively in the beekeeping industry as wood for the cover and also for the bottom board of beehives. It is a wood which is resistant to rot and deterioration and also is resistant to insects and the various ways in which they damage wood. However, we have never heard of an instance where redwood was harmful to bees, even when used on the bottom board. Consequently, we do not think you need to paint the inside of the bottom board although this would further protect the wood from deterioration.

DIVISION BOARD FEEDERS

Q. I purchased 10 of the Plastic Division Board feeders and am having problems. Floats were not included. I cut floats out of ¾ inch pine. The first ones I cut went about half way down and then became stuck on the edges of the four little embossed lines that are built into the feeder apparently for structural strength. The bees then proceeded to go around the end of the float and many of them were drowned. I then cut a float smaller in width so that as the bees took the syrup it would go beyond the embossed lines and go completely to the bottom. This wasn't good either as the float was too wide and again many bees drowned. These plastic division board feeders are tapered both in width and length. In addition the plastic is so smooth that a bee floating in the syrup cannot climb up the side.

A. Your problem with floats in the tapered plastic division board feeders is not unique. Others have had similar problems. The taper, of course, was put in there to allow the feeders to nest for shipping, to keep freight charges down. However, they do make it difficult for float purposes. One thing you might try is using either styrofoam chips, or perhaps some of the perlite that would be available through nursery or garden centers. This is the small styrofoam beads that are used to loosen up heavy soil.

QUEEN INTRODUCTION WITH A DOUBLE SCREEN

Q. I am not successful in introducing queens. In an earlier answer you suggested the use of a double screen. I have already made three single wire screens. I know the double screens are on the market but why double? And if so, what space should the screens be apart? Would 3/8 inch be enough?

I have a hive which is weak and would like to build it up as quickly as possible and at the same time introduce a new queen. I use two deep hive bodies the year around with shallow supers above for surplus honey. I suggest the following: Take a deep hive body and a frame or two of capped brood from a strong hive and two frames of honey, and fill up the rest of the body with drawn combs.

I would take off the cover of the weak hive which now has two shallow supers on and put on the screen with the entrance at the back. Put a wire screen over the entrance for a day or two. Then take the candy out of the shipping cage and put it in the hive, including the five or six bees that come with the queen. This would be a help to the queen until the brood hatched out. If all went well, I suggest leaving the screen on for at least a month and in this time the queen would get the smell of the hive below and both queens would be working in the meantime. I would appreciate your advice on this matter.

A. We would think 1/4 or 3/8 inch space between the screens would be sufficient but generally about 1/4 inch is used. This prevents some communication between bees in the lower and upper hive while still making use of the ventilation or heat from the bottom hive.

Your plan should work except we probably would use about four frames of brood — two with young brood and eggs and two of sealed brood or better yet, just emerging brood. When making up divides in the same yard, where older bees will return to the original colony site, it is better to make up the new unit on top. We always release the five or six attendant bees which come with the queen as it seems to increase the chances for success in requeening. This should all be done over a double screen so you have the combined efforts of the two queens. Remove the screen after a month.

EXTRACTOR

Q. I have five colonies of bees, but do not get enough honey to use an extractor. Last year I put the honey and comb in a double boiler, heating and crushing the comb while stirring, not allowing the honey to get over 150° F., but the honey never comes out clear.

Would you be kind enough to help me get clearer honey? I was wondering if the honey comb could be crushed in cheesecloth, without heating and let drain at room temperature.

A. You would get a better quality honey if you crushed the comb and allowed it to drain through cheesecloth in a warm room. We don't believe that you can afford to get along without a small extractor. It wouldn't take many sessions of melting or crushing combs to pay for the extractor, if only the cost of the foundation were taken into consideration. Save yourself a lot of mess and get larger amounts of better quality honey by obtaining an extractor.

EXTRACTOR SPEED

Q. I have a two-frame extractor that I operated by hand until this year when I put a motor on it. The problem is that I do not know if it is run at the proper speed. Could you tell me the speed at which a two-frame extractor should be operated?

A. The extractor should operate at about 300 RPM: you, of course, start at a lower speed so you can get a portion of the honey out of one side first, then reverse.

FOUNDATION

Q. I had always assumed that foundation was only made in one cell size but recently I have heard of the larger milled cell size foundation? I would like to have some to experiment with.

A. In the past, we have manufactured some foundation for the Indian honey bee, *Apis Cerana.* This foundation runs 1243 cells per square decimeter, considerably smaller in size than normal worker foundation which is 857 cells per square decimeter. Normally available would be 706 foundation, as well as drone foundation (520 cells per square decimeter). Many persons have found the 706 size especially suitable for cut comb production. Being intermediate in size between worker and drone, queens are reluctant to lay in it.

FOUNDATION

Q. What is the best way to nail foundation into the frames? Should the nails go in from the side or from the bottom to the top?

A. Put the nails in the top bar at a slight angle from the bottom to the top. This holds the strip of wood tightly against the foundation and frame.

ONE OR TWO HIVE BODIES

Q. Should each stand of bees have more than one hive body? Wouldn't more than two make it difficult to find the old queen?

A. Most colonies of bees do better if they have two hive bodies for the brood chamber. For honey storage, either additional hive bodies or shallow supers are used.

It is true that the more units there are in a hive, the more difficult it is to find the queen. Many beekeepers use a queen excluder between the brood chambers and the supers to keep the queen in the brood chamber only.

HONEY HANDLING

Q. I am planning to set up a small honey handling system which will enable me to pump my honey from the extractor instead of bucketing it. Since I consider myself a fair tinsmith, etc., I plan to use mostly materials at hand. According to my plans the honey will come in contact with, or pass through plastic pipe, copper, and, or, galvanized metal. Will any of these materials have a toxic effect on the honey or impair the taste or odor?

A. Almost all metals except stainless steel will affect honey to a degree. Coating the metal with a special lacquer which can be obtained from most bee supply houses will limit the reaction. Plastic pipe should be satisfactory.

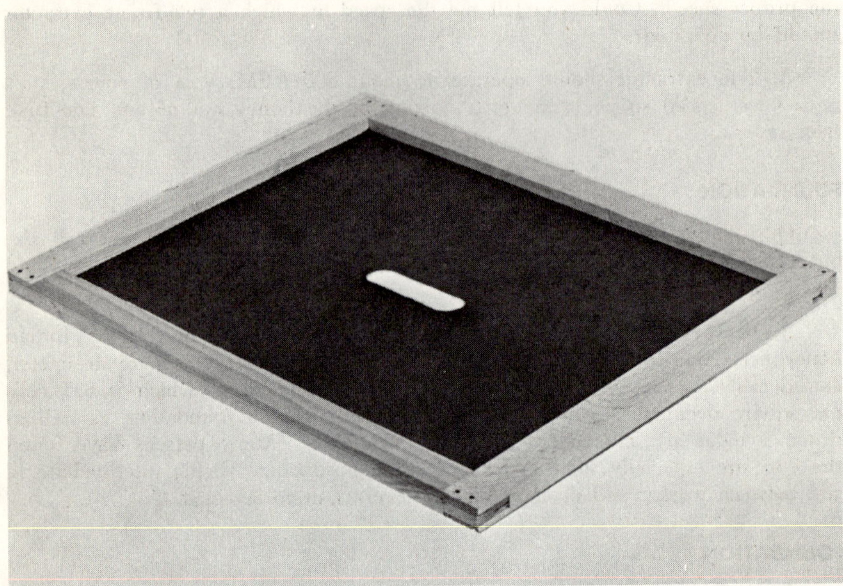

Fig. 46. The inner cover is a versatile item of equipment. It may be used for ventilation, equipped with a bee escape for removing honey, and is also useful in feeding.

INNER COVERS AND BEE ESCAPES

Q. I purchased a single hive of Midnight package bees last spring essentially to help the pollination of the several apple, peach, and pear trees and grape, barberry, almond, and filbert shrubs which I have on my lot. Should the hole in the inner cover be covered up in the winter? And which is the top and bottom of the inner cover?

A. The hole in the middle of the inner cover is put there for ventilation purposes as well as to allow the inner cover to double as an escape board. A bee escape, a small metal device that works like a gate, allowing only one way traffic through it, is inserted in the hole in the inner cover. The inner cover then is placed beneath the honey supers to allow the bees to escape down to the brood chamber. As the bees cannot return through the bee escape, the honey super will be comparatively free of bees by the next morning.

Many beekeepers also use some kind of a channel device to put over the hole in the inner cover leading to the outside so they have a "guarded" upper entrance for winter. When we use the inner cover here, we normally use the shallow edge of the inner cover down next to the frames. This allows less space for the bees to build burr comb if they should get crowded. Also, the deeper rim on top allows us to feed dry sugar to the bees on the top of the inner cover if we choose to do so.

PAINTING

Q. I am just getting started in beekeeping and have a question. I am thinking of painting the inside of the hive bodies, and am wondering if there would be any adverse effect on the bee colonies from doing this.

The purpose of the paint would be to prevent warping. Since there are many types of paint available, are there any that should be avoided?

A. Most references advise against the painting of the inside of the hive bodies. One of the reasons given is that the moisture which accumulates within the hive, is more apt to condense on the walls of the hive and continue to be too humid. The unpainted sides of the hive bodies can absorb some of the moisture and thus make the inside somewhat dryer.

There is one type of paint which you should very definitely not use inside the hive and that is a creosote base paint. It is all right to use on the outside or on the bottom side of the bottom boards or on the lids, but the odor acts as a repellent to bees, so therefore should not be used inside.

QUEEN EXCLUDER

Q. When does one put on queen excluders?

A. These are usually put on when the examination and reversing of the hive bodies is done. Inspect your bees, reverse the hive bodies, reduce the brood to the amount you want in the hive, put on the queen excluder and a super with drawn comb for extracted honey. The bees will store honey in the super during apple blossom time if the hive bodies are filled. Then when the sweet clover nectar flow starts, if you want comb honey, put the foundation right above the brood nest, between the body and the super. Move that super up and put a second super under it when the first is pretty well filled.

QUEEN EXCLUDER

Q. I have a hive of bees with a double brood chamber. On top of the second chamber, I have put a queen excluder. After the excluder, there are three shallow supers. Should this excluder be kept on the hive during the winter? If I remove it, will the queen (she's new) lay in the shallow supers? If she should get into the top shallow supers, what could I do about it?

A. Remove the queen excluder in the fall. The real danger of leaving the excluder on the hive all winter is that should the cluster move through the excluder in search of food, the queen would have to stay below the excluder where she would die from lack of food and the cold. The queen has tapered off considerably in her brood rearing at this time, and would not be so apt to go up into the shallow supers. This would be particularly true if the fall flow were very intense, and the bees were storing nectar up there.

QUEEN EXCLUDER

Q. It has always seemed to me that queen excluders are a hindrance to the bees. Would your shallow supers, in place of excluders, be just as effective in keeping the queen from entering the extracting supers during the season, or must one always use excluders to keep the queen down in the brood nest so she cannot get up into the supers above? If a shallow extracting super will keep her down in the brood nest will this result in greater honey storage?

A. The queen excluder is probably the most controversial item of equipment in all of beekeeping. The queen excluder is the only sure-fire method of confining the queen to the brood chambers and thus, keeping egg-laying and pollen storage out of the supers. This is important for comb honey production. If you are contemplating using a queen excluder, the welded wire excluder is probably the most effective and the easiest for the bees to get through, however, all excluders can easily become clogged and they must be kept clean.

An experienced beekeeper can determine supering needs at a glance and super to keep the bees working as he wishes. However, an inexperienced beekeeper may easily over-super and find that the brood nest has moved up into the shallow supers and this is why the queen excluder is recommended for many beginning beekeepers.

On the other hand, one of our experienced beekeepers had this to say:

In our experience the 6⅝" shallow super is better than an excluder, particularly in the production of extracted honey. However, in putting on supers, always super at the top above the supers where the bees are working vigorously. Should the top super be finished and sealed, then you'll have to go down until you get to where the bees are working hard and put your new super on there. In other words don't put the empties down next to the brood nest after the first supering or, of course, the queen will go into them. But when you top super this does not happen except in a very poor season and as you say, this plan without an excluder is inducive to greater honey storage.

Fig. 47. Diagram showing the method of supering. The supers are numbered in the order in which they are given to the bees.

RECONDITIONING USED EQUIPMENT

Q. We recently purchased a quantity of used bee equipment including hive bodies, supers, bottoms, lids and some very good frames. We would like to know how we can be sure they are clean. Can we boil the frames and excluders in lye water? We have scraped the supers and bodies clean and used a blowtorch on them. Is that enough?

A. You have correctly handled the used bodies and supers which you purchased by scraping them clean and using a blowtorch inside. The frames can be boiled as you have mentioned to sterilize them using about one can of lye to 5 gallons of boiling water.

SHALLOW SUPERS FOR EXTRACTING

Q. Why is it that small, as well as big honey producers, have gone into shallow supers for extracting? Shallow supers are almost unknown here in Argentina and are supposed to bring only double handling with no advantages. Colonies are worked here in 3 full depth bodies as a rule. Honeyflows are generally slow and very long, but in good years there are 3 separated "rush" periods in a season in which bees can fill a full depth super in a week.

A. There are several reasons for using shallow supers for surplus honey, but here are the two reasons heard most often: 1) the supers are filled more completely and sealed more quickly, and 2) they are lighter in weight to handle. These are good reasons, especially if you have light honeyflows and like to control the space more closely for storage. It takes a strong back to handle full-depth supers all day.

In Argentina and in certain parts of the United States, it would be much better and less work to use full-depth supers, since you would have fewer supers and frames to handle. A large number of our commercial beekeepers use the full-depth supers exclusively.

SOLAR WAX EXTRACTOR

Q. Will a solar wax melter be worth the investment?

A. During the hot months of the year a solar wax melter can help you save the wax that is in all those scrapings from the hives, the pieces of burr comb the bees manage to find places for if they get a little crowded, and the broken and poor combs that you discard. So often there is such a small bit of material at any one time that it is neglected or burned up just to get rid of it. Yet these same little bits, when thrown into a solar melter setting at the edge of the beeyard or beside the honey house, can render a surprising total amount of wax.

There is very little work involved. Just lift the glass top and throw in the material to be melted. The sun does the work for you. When the wax pan gets full enough, empty it when the wax is congealed. Occasionally the slum (the nonwax residue) must be scraped out too.

Solar wax melters may be purchased from most bee supply dealers and if you are handy at woodworking, they are easily constructed to any size that would handle the amount of material you have.

USE OF THE SMOKER

Q. I'm having real trouble using the smoker and examining the hive. I start out pretty well and then usually the smoker goes out, the bees get excited, and my orderly plan for examining the hive falls apart.

A. The smoker is a very essential tool, one which many beginners use improperly. Many materials are satisfactory for fuel, as long as they are easily available, easy to light, porous enough to allow sufficient draft so the fire doesn't go out every time you set the smoker down, and as long as the burning fuel does not smell so horrible that neither you nor the bees can stay in the beeyard. We use mostly old burlap bags even though they don't smell as nice as pine needles.

Fig. 48. A few puffs of smoke at the entrance to the colony sounds the alarm and the bees rush to open cells of honey to gorge themselves.

You must keep enough fuel in the smoker so it is really a smudge pot, not a fire pot — so that the smoke is cool and not hot. Hot smoke usually accomplishes just the opposite of what you want: instead of calming the bees by causing them to stick their heads in cells of honey to fill their honeysacs, the hot smoke causes them to become angry. Excessive and unnecessary pumping of the bellows creates a hotter fire, and in turn, a hotter smoke. This also helps to shorten the life of the smoker.

When you have finished using the smoker, the best thing to do is to empty the fuel and coals onto the ground and pour water on them until you are certain no fire is left. In case this is not advisable, both the smoke nozzle and the air intake hole at the bottom should be stuffed with green grass or leaves to suffocate the fire. A couple of corks to fit these holes are good for this purpose. One reminder — a smoker which is not emptied or properly stuffed can become a real fire hazard.

How much smoke should be used? There is no one answer. Generally, less smoke is needed on warm days, probably because propolis becomes softer and there is less popping and cracking when opening the hive, and because the bees are more apt to be collecting pollen, water, or nectar.

Fig. 49. Smoke is puffed over the top of the super and the hive tool is used to pry apart the super from the hive body. Smoke gently as the super is raised and try to make all movements slow and steady.

When opening the hive in cool weather, a few puffs of smoke directly into the entrance first thing gives you the jump on the guard bees. Then if the lid and inner cover are stuck with propolis, use your hive tool to pry as gently as you can. As soon as you have just a little opening, puff in a little smoke. Pry a little, puff a little — an occasional puff or two at the entrance. These recurring smoke puffs are like a boxer with a good left jab, continually pecking away, keeping his opponent from getting set.

With the inner cover off, the next step is usually to take out a side comb with a few or no bees on it. This allows the other combs to be pried sideways as they were loosened with the hive tool. Then, as they are lifted up, the bees on the two adjoining combs are not "rolled" or mashed together, thus minimizing the disturbance (and the chance of injuring the queen). As you become better acquainted with how bees react to intruders and to smoke, you will develop a "sensing" of their mood. From this, you anticipate a need so that you "jab" them in the face with some smoke just before they need it. Until this has become well learned, it is better to puff the smoke a little too often, rather than not enough.

Then, each individual frame with honey in it can be lifted out, bees and all, and then the bees can be brushed off in front of the colony using a bee brush. Now it would be nice if we could guarantee you that these methods will always work and that you will be able to examine your bees under any conditions without overly disturbing them but such proper use of your smoker will go a long way in the right direction. At first wear a veil and bee gloves and tie a string or rubber band around trouser legs or tuck the legs inside your socks. It is also advisable to work with the bees in a rather slow manner with even movements. Fast, jerky movements will have more of a tendency to aggravate the bees. Likewise, always work the colony from the side as standing in front of the entrance, or approaching a colony from the front, makes you a direct target.

SMOKER FUEL

Q. What is the best smoker fuel you have encountered?

A. Nearly every fuel has been tried from red peppers to tobacco. There is always something new for smoker fuel but many beekeepers return to using overalls or blue denim. It doesn't smell unpleasant, calms the bees, is long-burning, and produces a great cloud of smoke. Should there be any doubt just try a roll about the length of the smoker tank and about three or four inches in diameter. Insert the lighted end of the roll of denim in the smoker, keep the fire going with an occasional puff on the bellows, and you will find denim most satisfactory.

CLEANING A SMOKER

Q. I have used the same smoker for two years. It is very black and sticky inside and it is very hard to get anything to burn in it. What is the best way to clean the smoker and is there anyway to prevent the smoker from becoming black again?

A. The type of fuel you use will determine the amount of black deposit in your smoker. Fuels such as corn cobs, rotten apple wood and similar woods low in pitch content, will burn without creating the problem. The best way to clean out the black deposit is to scrape it out as much as possible and then wipe the insides down with a rough textured cloth. The grate should be removed and scraped. The air holes in the grate should be opened up and a rat-tail file is good for this. If you repeat the treatment as needed, you should be able to keep your smoker going for years.

SMOKER FUEL

Q. I am fairly new with bees, having worked them for about a year. One of my problems is that I can't seem to keep my smoker lit, and I'm sure it's me and not the smoker. I have tried fuel rags, denim, straw, burlap, cardboard and many other materials. It seems they all work well until I need the smoker and then it's out. I always clean the soot out before using, but somehow, I'm doing something wrong. What's the trick to keeping the smoker going?

A. The basic material used here for smoker fuel is burlap. We would guess that you are getting the smoker lit, and just getting it to smoking good, then going directly to the colony. We found out the hard way that you have to get the smoker going good and keep it puffing for quite a while before it will continue to maintain itself. If you are unable to do that, then frequent puffs of the smoker even if it's directed away from the hive will be a great advantage in keeping the smoker going. If you are operating only the one hive, you may find that the air-pressurized hive bomb may be more practical in the long run. Although it costs more than the conventional smoker for just one hive, it may work very well for you. It is liquid smoke that is in a pressurized can. We keep one of them handy as there are times when we want to run out quickly and take a look in a hive and not go to the bother of lighting a smoker.

CORRUGATED PAPER FOR SMOKER FUEL

Q. Quite sometime ago, you wrote of a recipe to soak newspaper in salt peter to light smokers. I've used it for many years but now the darn thing has got away from me. Can you find it?

A. Mr. Cale wrote: "I find that a roll or cylinder of corrugated paper soaked in salt peter solution with coloring matter and then dried, makes a tinder that cannot be put out. Just the touch of a match and it completely lights the smoker. Then other fuels may be added." The coloring matter is to mark the end which is to be lighted.

STRAW SKEP

Q. This year I decided it would prove interesting if, in addition to keeping bees in standard hives, I kept a skep of bees in the backyard for my enjoyment. How would one go about installing a package of bees or a swarm in a straw skep?

A. Frankly, we have never done this but this is how we would go about it. We'll start with a package of bees. First of all we would prepare the skep by suspending one or more combs in it. It would be better to use combs that contained some stores of pollen and honey. First we would copiously feed the bees in the package by brushing the screen with sugar syrup (2 parts sugar to 1 part water). Then we would remove the queen cage from the package, remove the attendant bees from the queen cage, remove the paper from the candy end of the cage and make a hole through the candy with a wooden match to enable the bees to eat through and release her, and then suspend the queen cage in the top of the skep. We would then shake as many bees as possible from the package into the skep and turn the skep over on its stand, placing the package containing the balance of the bees in front of the entrance to the skep. It would be advisable to do this in late afternoon or evening so the bees become settled in the skep and do not fly out freely in the open air.

In the case of a swarm, one would need to find the queen, cage her, and then suspend the cage in the top of the skep. The swarm then should be shaken into the skep or as many bees as possible brushed into it, and the skep turned over onto a stand near where the swarm settled. The remaining bees in the swarm will join the queen in the skep and we believe they would stay and build a colony there.

SUPERS

Q. I'm just starting beekeeping, I have some old supers. A man told me I have to wash them in salt water. Another told me I have to char them all. Which should I do?

A. The only reason for making an attempt to clean up those old supers in the first place is if you suspect that they may be contaminated with the disease organisms from the foulbroods. Washing in salt water would very likely have no effect, although the suggestion to char the inside of the super shells would be effective.

To be doubly safe, preventive treatment of doses of Terramycin might also be suggested, as long as feeding of the drugs was discontinued thirty days before the start of the honeyflow.

VENTILATING BLOCKS

Q. What are ventilating blocks and where are they used?

A. Ventilating blocks are small blocks of wood about one inch square, which some beekeepers place at the front of the hive between the hive body and the bottom board to give the bees more ventilation. This is supposed to help prevent swarming, especially for comb honey production. These should only be used in periods of hot weather when the bees are in real need of help in keeping the interior of the hive at workable temperatures. They should be removed in cooler weather and when there are signs of robbing activity.

WHITE CLOTHING FOR BEEKEEPERS

Q. Do beekeepers wear white because bees recognize this color?

A. It is true that beekeepers wear white clothing to enable them to work their colonies of honey bees without being stung to any extent. This is because the compound eye of the honey bee recognizes the movement of a dark object much quickly than one which is light colored or white. This movement of a

Fig. 50. This beekeeper is fully equipped. He may later discard the gloves as he gains experience in handling bees but the veil will remain a standard item as will the practice of tying trouser legs.

dark object excites the bees to protect themselves and this often results in stinging. For example, a beekeeper working bees will often remove a wrist watch because angry bees are inclined to head toward it. In a similar way, they will fly toward the eyes or the dark opening of a mouth in much a similar way. This is why beekeepers wear bee veils. Similarly a beekeeper wearing white socks will not be stung as often around the ankles as would a beekeeper wearing black socks.

The vision of the honey bee is quite different from that of man. The colors red and yellow at that end of the spectrum are seen as greys and even black objects and, therefore, the bee is color blind with respect to these colors. The bee does see the greens, the blues and the violet colors and then continues to see in the ultraviolet range where man does not see. Colin G. Butler, in his book entitled "The World of the Honeybee" states, "Some white objects absorb ultra-violet light while others reflect it. This means, of course, that two objects which appear equally white to us may appear as of two quite distinct colors to bees — the first can readily be distinguished by them probably as a shade of blue-green, whereas the other kind of white object probably appears as a shade of white." He continues, "Therefore, since most white flowers absorb ultra-violet light, such flowers probably appear to be blue-green to bees."

We do not know whether the white cloth of a uniform for beekeepers would reflect ultra-violet light or not but have been informed that ultra-violet light is used to determine the whiteness of clothing in studies of various detergents used for cleaning them. It is claimed that the use of ultra-violet light enables one to tell whiteness better than when ordinary light is used. Who knows, perhaps the white uniform of a beekeeper might also appear blue-green to bees.

WOOD PRESERVATIVES

Q. 1. Is Pentachlorophenol (Penta) wood preservative harmful to bees if thinned with fuel oil and painted on tops and bottom boards?

2. Is Penta harmful to bees if thinned with paint thinner and painted over?

3. If thinned with paint thinner and unpainted?

4. We are supposed to have a standardized beehive, yet some advertise hives as 19⅞" long and some advertise hives as 20 inches long. This is a very short difference but a most important one. Which is standard and which size do you recommend?

A. 1. Pentachlorophenol wood preservative, thinned with fuel oil and painted on tops and bottom boards and allowed to soak into the wood and aired to some extent, is not harmful to bees. We have been treating bottom boards and outer cover rims with this material for many years and have received no word that there was harm at any time.

2. We do not know what paint thinner you will use to thin the Penta but if painted over with ordinary beehive paint, it would certainly not be harmful to bees.

3. Without knowing what paint thinner was used with the Penta, we would not be positive of our answer here but fully believe that if allowed to soak into the wood and aired to a certain extent, it would not be harmful to bees.

4. The major suppliers manufacture a hive body 19⅞" long and we are quite confident that this is the standard length. Naturally, we would recommend this length.

FEEDING

EARLY SPRING FEEDING

Q. Will feeding in early spring stimulate brood rearing?

A. By starting to feed the bees the latter part of February, you can stimulate brood rearing. In fact, too much feeding might cause the bees to build up so fast that they are ready to swarm by fruit blossom time. While feeding, be sure to use antibiotics in the feed to prevent disease and also stimulate brood rearing. If you do feed early, be ready to care for your bees early in order to prevent swarming. Bees fed early will build up rapidly and give you an abundance of brood in case you want to make some early divisions in the spring.

FEEDING FOR SPRING BUILDUP

Q. Please tell me the proper date to start feeding my colonies of bees in order to make them strong for the late spring honeyflow.

A. A good general rule would be to start feeding at least six weeks ahead of the honeyflow. The timing of the honeyflow will vary in different parts of the country but if your honeyflow (Nebraska) begins in early June, as ours does, then you should start to feed by the middle of April. The most important thing to remember is that under no conditions should the colony be allowed to run short of food during this crucial period of spring buildup.

STIMULATIVE FEEDING IN APRIL-MAY

Q. For stimulative feeding of bees in late April and May, just how much syrup should be fed in quarts, gallons, etc.? I am using pails with punched lids.

Fig. 51. These friction-top pails have holes punched in the lids so the bees can take up the syrup as they need it. An empty super fits down over the pails and protects them. This system is generally acknowledged to be the best for feeding colonies.

A. If you are just feeding for stimulative purposes, it would be better to feed the bees warm syrup. This syrup should be two parts sugar to one part water. You would need to feed at least a couple of pails to get the queen into heavy brood production. You would want to start this feeding if possible, six weeks before your honeyflow. This would provide a large colony by the time that the honeyflow hits.

FEEDING SUGAR WATER

Q. I got a book with my bees that told how to handle them. In the book it said to feed the bees sugar water until they would not take it anymore. I have been feeding them sugar water since the middle of April although the brood hive is full and one super is full of honey and I have put a second one on the hive. Still they eat the sugar water — can I stop now?

A. Your bees will continue to eat sugar syrup as long as you feed it, especially if it is of a higher sugar content than the available nectar. You should certainly stop feeding them the sugar. We usually feed until the brood chamber has four or five frames of honey.

FEEDING WEAK COLONIES

Q. We have three colonies, all caught at swarming time. Two of them had two full supers for brood and food, while one that we caught late in the season only filled one ten-frame body. Will feeding be necessary?

A. The colony in a single ten-frame body undoubtedly will need feeding. Our source tells us that bees at the close of brood rearing in the fall should cover 20 or more combs, have 45 pounds of honey in the top chamber in dark brood combs and 15 to 30 pounds of honey in the lower chamber. If it is not possible for you to establish these conditions, watch closely and feed the colony from time to time as required.

FEEDING BETWEEN HONEYFLOWS

Q. What proportion or volume of water should be mixed with honey to stimulate bees to draw foundation during a dearth of nectar between honeyflows?

A. We would feed the honey without dilution because, if you diluted honey with water to any extent, you would be faced with fermentation and fermented honey is not good bee feed. If you use sugar syrup, you should feed a heavy syrup made with two parts of sugar by weight to one part of water.

FEEDING IN LATE SUMMER

Q. Although it is now Aug. 15, I find very little honey stored in any of the hives. The bees have been working apparently. Each colony has increased tremendously in numbers but they do not seem to have produced much honey. We had a cool, wet spring and early summer here, this condition lasting until most of the clover bloom was gone. This is probably the reason for no honey. Would it help to feed these bees now or should I wait until fall and see if they can gather enough to winter themselves through?

A. You should probably wait until after a killing frost before feeding the bees. Give each colony 20 to 25 pounds of sugar made into syrup if they have not stored any honey. Use 1½ parts sugar to 1 part water by volume.

FEEDING FOR WINTER

Q. I will have to feed my bees to bring them through the winter. I want to know how I am to determine when to start feeding? Since I plan to winter them outdoors, how will I be able to place the feeders in the hives in cold weather?

A. Feeding in the fall can be started any time during the period following the end of your last honeyflow of the year. In your case we would strongly suggest that you start to feed as soon as possible.

You ask how you will be able to place the feeders in the hive in cold weather? Since you do not mention what kind of feeders you intend to use, we will try to make a note concerning several different types and hope that we hit upon the right one for you.

The Boardman feeder or the entrance feeder that consists of a block of wood in which a fruit jar sits holding the syrup, is of little or no value to use at this time of year. The bees cannot get down to this source of feed during cold weather.

If you are intending to use a division board feeder, the kind that hangs in the hive body similar to a frame, this should be placed next to the first comb which has bees on it. If it is left at the side of the hive, it will be difficult for the bees to get over to it in cool weather.

The preferable method is to use 10-lb. tin pails, 5-lb jars, or 7-lb. plastic pails to feed the bees. The lids on the buckets and jars can be punched with a 1/16" hole and the plastic pail lids are better if the 1/16" hole is drilled in them. If your colony needs no more than two buckets of feed to give it sufficient stores to winter, then these may be placed directly over the hole in the inner cover providing the bees are clustered directly under it.

If the bees are requiring more than two buckets of feed or are clustered somewhere other than under the hole in the inner cover, then it would be better to remove the inner cover and put the buckets of feed directly over the cluster of bees.

If it is necessary to feed in such a manner, the inner cover could be placed on top of the hive body which must be placed over the original brood chamber to provide room for the feeders. The hive cover is then placed on top of this empty body.

If it becomes necessary to leave the feed on into extremely cold weather, it might be well to pack burlap bags or some kind of insulation around the pails of syrup and to fill this empty space in the feeder body.

FEEDING FOR WINTER

Q. I keep 6-8 hives of bees in a highly populated area where the lot size is about ¼ acre. I can throw a rock in any direction and hit a house. I say this so you will understand when I say we have no plants to produce a fall honeyflow.

The honey has been removed and I'm ready to winter the bees. In the two brood chambers I find plenty of bees, good brood nest and a fair amount of pollen but very little honey. How do I get more honey for wintering? If I leave a super of honey on the hive then next spring I'll have the brood in this super. What do I do?

A. You could feed them a sugar solution and the best way to feed in the fall would be two parts sugar to one part water and warming it while you mix the solution. You would need to buy friction top cans and poke holes in the tops, fill them with the sugar solution, and each pail should be inverted and placed over the brood area.

If you do not want to feed the sugar solution, you could go ahead and let the bees have that other super of honey. They would move up in the super as you suggested; however, in the spring the bees would move the queen down below when the honeyflow starts.

FEEDING IN WINTER

Q. I have been told not to break the winter cluster for feeding purposes and I can understand that but how do I tell if the bees are in need of winter feeding to stave off starvation?

A. If, following a snow, you notice some colonies on which the snow on the cover starts melting sooner or more so than on others, you should mark that colony or those colonies to be checked for food. The cluster has probably eaten its way to the combs directly under the lid, and their heat has melted the snow. Those with plenty of honey cluster lower in the hive and the snow will remain on the outer cover.

Much winter loss occurs from starvation. The best and cheapest cure to prevent it is by leaving plenty of honey in the fall. Weighing a colony doesn't give the entire picture — it's the location of the honey that really counts. Starvation may come in very cold weather from having lots of honey on both sides of the hive with very little in the center.

As honey is consumed, it is normal and easy for the bees to move up between the combs. When it is cold, the cluster finds it difficult to move over the tops of combs, or around the ends to reach the full combs. Should the cluster reach the top of the combs or run out of honey in the combs they cover and have a period of cold weather, the chances are good they will starve, perhaps in the midst of plenty.

Obviously the risk of starvation increases as spring approaches. Once the colony has renewed brood rearing, the honey supply dwindles fast. Besides feeding themselves and the brood, the bees must maintain a brood-rearing temperature of about 93° F. in that area of the cluster and the source of energy for that heat will have to be winter stores.

On real nice days in winter and very early spring, it is possible to switch full combs for empties or to feed. A tin pail or a jar with several small holes punched in the lid is inverted directly over the cluster, making sure the holes in the lid are between the frames so the bees have access to the feed. An empty super protects the feeding container.

CRYSTALLIZATION OF SUGAR SYRUP

Q. We are having trouble with sugar syrup crystallizing in the boardman feeder. We use one cup of water to two cups of sugar. The water is heated to the boiling point before stirring in the sugar. I usually prepare a gallon of syrup at a time. The syrup will not crystallize as long as it is in a glass jar in the cupboard — no refrigeration. As soon as I pour it into the boardman feeder it begins to harden around the mouth and neck of the jar. In 48 hours all the holes are plugged. What is wrong?

A. You can probably eliminate at least part of the problem of crystallizing sugar syrup by reducing the amount of sugar. Try using 1 part water to 1½ parts sugar. You should also add a small amount of cream of tartar. (1 teaspoon per 25 pounds sugar.) This helps to retard crystallization. You might also try using a 5 pound pail with 6 or 8 small holes in the lid. Invert this pail of syrup over the hole in the inner cover and protect it with an empty shallow super.

FEEDING DRUGS IN SYRUP

Q. I have never fed my bees powdered sugar but have made a simple syrup with granulated sugar. I have had considerable trouble, at times, getting the drugs to mix properly in solution. I would appreciate your comments on the proper mixing procedure for drugs.

A. Your statement about the difficulty of incorporating drugs in syrup solutions is quite valid. Terramycin shouldn't be used in sugar syrup because it breaks down and loses strength in a short time if fed in this manner. We should also mention that there is the problem that such drug containing syrup might be stored in honey combs and not consumed, then becoming a danger later when extracting the crop. Of course, no drug feedings are recommended prior to the honeyflow.

FEEDING FULLER CANDY

Q. I have just read about "Fuller Candy" for feeding bees. It is supposed to be so much better than feeding sugar syrup. NO chance of encouraging robbing and is better to handle, etc.

Can you tell me about this and how to use it and where to get it?

A. The Fuller Candy recipe makes a soft fondant, similar to that in chocolate creams. It may be made to the degree of hardness or softness one prefers. The higher the temperature at which the candy is boiled, the harder it will become.

INGREDIENTS:

12 pounds granulated sugar
1½ pounds liquid glucose
1¼ quarts water (5 cups)
¼ teaspoonful cream of tartar, added when temperature reaches about 230° F. (110° C.)
Boil to 238° F. (114.4° C.)

As soon as the sugar begins to dissolve, prior to boiling, the spoon or paddle used in stirring should be removed from the kettle. Do not stir the candy while cooking, as this will cause a coarse grain. Boil to 238° F. (114.4° C.).

While the material is cooling to about 125° F. (51.6° C.) stir vigorously until the mass appears in color and consistency to be like boiled starch. At once pour into molds or feeders and cool.

FEEDING SUGAR FOR WINTER STORES

Q. I am wondering about the economics of feeding sugar or honey to bees to insure a good supply of winter stores. Can I get the same effects feeding half as much honey as sugar? Would the honey (extracted) be best fed in a dilute solution of say, 2 gallons feed or undiluted? Would the addition of terramycin to the feed be sufficient insurance against spread of AFB. In using sugar syrup, would the high percentage of sucrose be detrimental to the health of the colony?

A. In relation to feeding bees, there are several things to be considered. For example, if a hive needs 40 pounds total weight of winter stores, then this would be best provided in pure natural honey rather than a honey syrup. If a deficiency in winter stores were to be supplied by feeding sugar syrup, however, then the total amount of sugar syrup fed would have to be sufficient to be reduced by evaporation to an amount which would be equal to 40 pounds of honey.

To feed honey in a diluted form is a bit risky especially if the weather is somewhat warm. The reason for this is that the diluted honey has a tendency to ferment quite rapidly and this fermented solution would not be good for bees to winter on.

The addition of terramycin to a syrup solution is not advised, since the terramycin breaks down quite quickly in a liquid feed. Use TM25 mixed 1:3 with powdered sugar as a dust treatment for American foulbrood.

When sugar syrup is made from cane or beet sugar, the proportion should be one part sugar to one part water if it is necessary to mix this syrup cold. If you are able to use steam to dissolve the sugar in the water, you can use two parts sugar to one part water. We prefer to use the more concentrated solution. If your colonies have enough pollen and are fed sufficient sugar syrup to insure that they will make it through the winter, generally speaking these colonies will survive quite well.

FEEDING POWDERED SUGAR

Q. Could you tell me anything on feeding powdered sugar as a syrup in the fall? Will it affect the bees in any way for wintering? I would like to feed a syrup using powdered sugar.

A. There is a very good reason why powdered sugar should not be fed to bees in any sizable quantity. The powdered sugar contains about 3% starch, and this starch causes a stoppage of the intestinal tract of the honey bee. It is actually the reverse of dysentery. The small quantities that bees get from queen cage candy or from sugar-drop dusting mixtures does not cause this, but the feeding of large quantities of sugar syrup made from powdered sugar could lead to problems.

Fig. 52. When bees fly freely and the weather is warm, it is often possible to feed dry sugar which may be spread on the inner cover, leaving the hole in the center of the cover open. Five pounds of sugar will last longer than 10 pounds of syrup, particularly when the bees have access to minor sources of nectar before the honeyflow.

FEEDING BROWN SUGAR

Q. Is brown sugar suitable as bee feed? I have a good source of bee sugar and frequently they have brown sugar.

A. Brown sugar is not suitable as a bee feed. This is particularly true in times of inclement weather or cold weather when the bees are confined to their hive. Use of the brown sugar would result in a dysentery of the honey bee and probably loss of the colony.

FEEDING HONEY

Q. Tell me how to have the bees eat their own honey. We have a good healthy hive that is rapidly growing but they are using more sugar and water than they did initially. I estimate the hive has grown from 7,000 to about 18,000 in 9 weeks.

A. In a sense, you have answered your own question concerning the use of the sugar syrup by the growing colony of bees. As the colony grows, the number of larva which are being fed to develop into more worker bees, continues to grow. This places quite a demand upon the existing bees to provide enough nectar and pollen to feed them properly. As long as there is insufficient nectar available from the flowers in the area, it is necessary that they be fed sugar syrup.

Usually, whenever the floral sources of the area provide enough nectar that the bees can work on this and get all of their living from the flowers, they will not take the sugar syrup from the feeder. In other words, whenever the bees have a choice, they prefer natural nectar from the flowers to the sugar syrup. One of the best beekeeping management practices that you can employ is to see to it that the colony of bees does not run short of food. This is the one situation which will slow down their development and prevent them from becoming large enough to provide a crop of honey not only for themselves but for you, also.

FEEDING HONEY

Q. I feed honey to a few colonies during dearth to get them to draw foundation. Lately, I fell heir to some honey of unknown ancestry and, to make it safe for feeding, could I mix it with terramycin, and in what proportions?

A. Feeding honey of unknown ancestry is often an open invitation to disease and more trouble disease-wise than any savings in the long run. If you must feed this honey, liquefy it if it has granulated, use terramycin as a preventative for AFB and check the colony closely for any signs of disease.

Terramycin loses its potency too quickly in liquid form to be effective as a preventative against American foulbrood disease. In addition to feeding the honey, we suggest using a dust treatment consisting of terramycin (TM25) mixed 1:3 with powdered sugar. Four teaspoonfuls of the dust should be applied over the brood combs two or three times in early spring at seven to ten day intervals.

FEEDING OLD HONEY TO BEES

Q. Will you please tell me if you think old honey, granulated and age unknown, will harm my bees? It smells like sour apples, but tastes like last year's honey (late flow.) This honey was stored in an old barn and exposed to the cold. I am sure the honey is several years old, maybe 15 or 20 years old. It was in glass jars.

A. The old honey will not hurt the bees but unless it is used only for brood rearing, it will affect the flavor of your honey.

FEEDING DARK HONEY

Q. I have some dark honey I want to feed back to the bees. How should I prepare it?

A. The honey that you have and want to feed back to bees can certainly be used, but care must be taken if you aren't certain of the origin of the honey. If it is your own, and you know that it is free of disease spores, there is no problem. If you have suspicions that disease may be present, you should use drugs included with the honey as a precaution. Follow instructions on the label for correct preventative dosages.

We assume that the honey is dark only because it is from floral sources that produce darker honey, not because it has been carmelized or scorched. Scorched honey can be a cause of severe dysentery for the bees. The honey should be fed full strength, rather than diluting it to feed back to the bees. The dilution will work all right, but the honey would ferment rather quickly.

FEEDING DRUGS

Q. When feeding bees drugs for the prevention of disease, can the drugs be mixed together and fed at one time or must it be done separately? I have some poor grade honey from last fall which I figured on feeding back to the bees in this way in the spring.

A. Follow the directions for each drug but yes, most can be mixed together for feeding. It is wise to add the drugs to honey which is being fed back to the bees, especially if you are not sure of the source of the honey. Always check the directions for each drug for specific uses. For example Terramycin should not be fed in liquid.

FEEDING "EXPELLER PROCESSED" SOYBEAN FLOUR

Q. What is meant by the term "expeller processed "soybean flour?

A. There are two processes generally used in the United States for removing oil from soybeans. The first, which is mainly used, is the extraction process which uses a chemical solvent to remove the oil. The second is a mechanical process which uses screw presses or expellers to express the oil from the soybeans. The products resulting from this type of processing are often referred to as "expeller," e.g., "expeller soybean meal" and "expeller processed" soybean flour.

FEEDING POLLEN

Q. In these tests where they feed bees pollen and show such good results do you think the good results from the pollen feeding is because the bees haven't had to eat their own honey and in the process of doing this, have not uncovered their own pollen. I am completely in the dark as to experience on this. I have never fed any pollen on the side, only as bees get their own pollen. Say in February, a hive of bees haven't uncovered any pollen where they are in the hives, but there is some sealed pollen on the other side of the hive. Now they need pollen to start raising brood. Will they pass over to the other sealed honey and get at this sealed honey or will they just wait until they get some natural honey or will they consume honey over to this other pollen.

Now, I am sorting out some pollen combs. I am a rebel beekeeper so to speak and I work usually without excluders. I have more pollen in honey supers on account of this. I'm thinking of putting some of these in a body above inner covers with an escape entrance opening. Will bees go up and get this? Pollen last year started to come on April 7. I plan on doing this about a month earlier.

A. The results are due to having the abundance of food necessary to stimulate brood rearing. If the bees failed to uncover stored pollen, it would prove only that they had plenty without using the stored pollen.

Your second question concerning the bees ability to detect pollen stored under honey on the other side of the hive. If the weather is quite cold, and the cluster tight, they undoubtedly would not make the trip over for the pollen. In addition, we seriously wonder whether they actually know that the pollen is stored under that honey. We suspect more likely that as they consume the honey, the pollen is uncovered and they then go ahead and use it.

The bees might use the pollen from the pollen combs placed above the inner cover. Better, put the pollen combs under the brood body(s). Bees prefer not to use old pollen if fresh pollen is available. Real old, dried pollen won't be used at all.

FEEDING POLLEN SUPPLEMENT

Q. I have never fed much pollen supplement and I don't know a good way. Some recommend mixing it with plain water and some syrup. I have tried dissolving a portion of pure pollen in warm water and then adding the mixture to the substitute, making it into cakes as per instruction. I trust it won't sour if made in advance. I am also wondering if it should be refrigerated?

A. There are a number of ways that pollen substitutes can be fed. Some feed it straight without the pollen mixed with it. The mixing of pollen into the material makes it not only more attractive but if the percentage of pollen is high enough, the bees can continue to rear brood in the complete absence of natural pollen. One method of feeding the dry pollen supplement is to place it in a container out in the beeyard. Some kind of a rain shelter should be put over the top. If the weather is warm enough, the bees will fly and gather it from this feeder. However, you might find it much to your advantage to feed it right over the cluster. This would then eliminate the question of poor weather. A rich sugar syrup solution should be used, about two parts sugar to one part of water. You will probably have to heat the solution before all the sugar will be dissolved. While this is heating, you might take the pollen substitute and mix the dry pollen thoroughly with it. Then, you can just add in the syrup and knead the mixture until it is a very thick paste. A patty of this could then be placed right on top of the top bars beneath the inner cover. We normally refrigerate the moistened mixture until used.

Fig. 53. A pollen cake being fed to a colony in early March. (Photo, Univ. of Wisc.)

Fig. 54. This colony has nearly consumed its pollen cake. The colony should be at full-strength by honeyflow.

FEEDING POLLEN SUBSTITUTE

Q. I have a question that needs answering and I thought you would be able to do it. We keep ground calf feed in a metal bin, not too far from my hives, and on warm days the bees flock down to the feed bin and begin hauling away something in it. Could you tell me what they are getting and if they need it or not? The feed consists of: 1) Ground ear corn, 2) Ground limestone (mineral), 3) Salt, 4) Phosphate (powder), 5 Aureomycin and 6) Vitamin A supplement.

A. The problem you had with your bees trying to gather ground feed on nice open days in early spring is not unusual. When bees are in need of pollen, such as is the case in early spring, they often will gather materials which they cannot use at all, even to the extent of gathering finely ground sawdust. The best way to eliminate the problem is to feed them a dry pollen substitute within the yard itself. Such pollen substitute is marketed by bee supply companies. The preferable way to feed it in these circumstances is in a box with some kind of a lid over the top to keep the rain out.

FEEDING ISOMEROSE

Q. There has been quite a lot of discussion lately about Isomerose corn syrup. Possibly it could be used as a desirable bee food. Would you enlighten us as to its general properties and uses.

A. Isomerose is the registered trade name of the Clinton Corn Processing Company for a high fructose corn syrup manufactured from edible corn starch. It is produced by an enzymatic process which involves liquefaction of the starch, followed by saccharification with glucoamylase, followed by isomerization of approximately half of the dextrose to fructose. The remaining saccharides are a mixture of simple and higher polsaccharides. The product has a sweetness, on the dry substance basis, comparable to sucrose. It is water white in color, is slightly viscous and is extremely bland in flavor, its only flavor being that of sweetness. It more nearly resembles invert sugar than any other nutritive sweetner on the market.

Isomerose 100 Brand High Fructose Corn Syrup is shipped at 71% solids and maintained at elevated temperatures. It is necessary to do this because dextrose will crystallize out of the solution unless the temperature is kept above 80° F. The product is used widely in food processing, in such products as carbonated beverages, still drinks, pickles, salad dressing, yeast raised baked goods and other food products. When used, it is declared as either glucose syrup or corn syrup in the ingredient listing on the package.

There is the possibility that Isomerose 100 Brand High Fructose Corn Syrup might be a desirable bee food. In those places where sucrose is dissolved, to be used as a bee food, it would seem to me that Isomerose 100 Brand High Fructose Corn Syrup could also be used. However, I am not knowledgeable in this area and this would be something that the beekeeper would have to evaluate. Bees love this product because whenever there is the slightest spillage at any time in the summer, a swarm of bees appears almost instantly.

Typical Analysis of Isomerose 100 Brand High Fructose Corn Syrup Compared With Average Extracted American Honey

	Isomerose 100	Honey*
moisture	29%	17.2%
solids	71%	82.8%

Dry Basis Comparison

ash	0.03%	0.205%
Carbohydrates:		
Dextrose	50%	37.8%
Fructose	42%	46.1%
other saccharides	8%	12.4%
Relative Sweetness	100	125

FEEDING TERRAMYCIN MIXTURE

Q. I note that the literature recommends that Terramycin be mixed with powdered sugar and sprinkled over the frames and that none recommends that it be mixed with syrup since it is claimed this tends to weaken the potency of the drug. If this is true, why does Pfizer recommend that the drug be mixed with water for the control of diseases in poultry?

A. It is true that we recommend the use of Terramycin as a dust rather than in syrup because in a liquid it soon loses its potency. While Terramycin can be used in syrup fed to bees, its potency disappears over a period of time. You will realize that this syrup may be in the feed can for a week or 10 days before the bees take it down and then it may be stored in the comb for longer periods of time. In contrast, when mixed with water daily in the control of disease of poultry, it is consumed quickly and does not have a chance to lose its potency.

HONEY

HONEY GATHERING

Q. How much honey can or does a honey bee gather in her life span?

A. A typical bee-load of nectar is 30 mg. and she may average 10 trips per day. Loads vary considerably according to temperature, honeyflow, and kind of plant being visited. Bees live an average to six weeks, but forage only during the last three weeks of life. Therefore, at 300 mg. a day for 21 days, a typical bee might gather a total of 3600 mg. (3.6 grams) of nectar. When concentrated into honey, this would be about 1¼ grams (0.0402 oz.)

SECTION HONEY

Q. Is there a market for comb honey in the 4¼ x 4¼ sections? Years ago I raised a lot of it and there was a market in the stores and then there was no market so we stopped producing it. I would like to go into the business, but am unable to find if there is a market. I think what happened to the market was that the chain self-service stores could not handle it. The customers would handle the sections and break the combs. For that reason they would have had to stop handling section comb honey.

A. Section comb honey is always a special product and we have found that the market today is extremely good for section honey because it is so rare. Many specialty stores, open air markets, health food stores, etc. are looking for producers to supply their customers. One way around the problem is to produce bulk comb honey and use the new plastic boxes specifically designed for the 4¼ x 4¼ cut of honey. The lids snap on tightly and the product is easily displayed with no fear of breakage.

HEATING OF HONEY

Q. Is it necessary to heat honey? I find that honey of my crop when unheated is much lighter than the heated honey of another beekeeper's crop produced the same year about 3 miles from my yard. His honey granulated just as quickly or slowly as mine, however you look at it.

A. Heating of honey can accomplish two things. The first of these is that heating honey has a tendency to kill any yeast organisms that may be present in the honey and thus prevent fermentation. The second thing that is accomplished by heating is that it has a tendency to liquefy any crystals and thus slow the granulation of honey. However, heating of honey will darken the honey and will to some extent, destroy some of the natural flavor and aroma of honey. If you can avoid heating we certainly would recommend that you do so. If honey has a moisture content of 18.2% or less, there is no chance that any yeast formation will occur anyway.

GRANULATION OF GOLDENROD HONEY

Q. My main honey crop (Ohio) comes from goldenrod in the fall. My customers are all well pleased with the taste and flavor. Now, the problem is the honey granulates and is returned to me from the grocer's shelves. What can I add or do to prevent this granulation? My mother used to add cream of tartar to taffy to keep it from sugaring out. If I state it on the jar, would I be safe with the Pure Food Administration? Another alternative that was proposed to me was to mix the goldenrod honey with some clover honey which I have on hand. If you favor this, what proportions should be followed?

A. We would discourage any additions such as cream of tartar to honey. Honey is a pure food and people want it that way. You would have to check with the Pure Food and Drug Administration on the matter but you might have labeling difficulties if any substance is added to the honey. Unfortunately, goldenrod honey does crystallize quickly. Heating retards the process. Cool the honey as quickly as possible. You should also pack only enough honey to meet the demands of your market.

Clover honey is slower to crystallize than goldenrod honey and a mixture of 50% might slow down the process but the attractiveness of the goldenrod honey would be diminished. You might try selling crystallized goldenrod honey with descriptions of ways to liquefy the honey. You might also investigate a honey spread. You may have to do some consumer education, but it's a cinch that goldenrod honey will keep better in the crystallized form than as a liquid.

SALVAGING FERMENTED HONEY

Q. I have a problem with fermented honey and wonder if there is anything that can be done to salvage it? Also does leaving it exposed to the air allow it to accumulate moisture so that it ferments or are there more probable causes.

A. The degree of spoilage or effect upon flavor and quality, depends upon the length of time fermentation has been allowed to proceed before being stopped by heating or some other treatment. If fermentation and subsequent spoilage has not proceeded too far, you can heat the honey to destroy the yeasts and place it in a sealed container which should stop all future fermentation. It is best to heat the honey in a closed system, heating it rapidly and then cooling it rapidly inasmuch as the least heat exposure results in the least damage to the honey. The common recommendation for control of fermentation is to heat the honey to 145° F. and then to cool it as rapidly as possible. If this is done it will be safe from fermentation if protected from further yeast contamination.

Leaving the honey exposed to air certainly allows it to accumulate moisture from the air inasmuch as honey is hygroscopic. The additional moisture would tend to add to your fermentation problem because honey of high moisture content is more subject to fermentation. Inasmuch as the soil and air, as well as equipment, are highly contaminated with honey tolerant yeasts, keeping the honey in a closed container, especially after a heating process, is essential.

HEATING OF HONEY

Q. Will you please let me know at what temperature the enzymes in honey are destroyed? The Dept. of Agriculture of California requires that extracted honey be strained through a silk or nylon cloth of very fine mesh and to do that, the honey in 5 gallon cans has to be heated. On the other hand, people are becoming more health conscious where honey is concerned and are demanding that it be unheated or at least not heated to a degree that will destroy the enzymes.

No doubt the large commercial producers of bottled honey have some mechanical way to strain honey without heating. I have seen bottled honey in a health food store labeled, "Unblended and uncooked," which doesn't say that it has not been heated to some degree, merely that it has not been cooked.

Can you explain how the small beekeeper can produce extracted honey that has not been subjected to heat to the point of damage? Without heat it wouldn't go through the strainer any more than cold molasses would. We take off our honey in July, but don't need to strain it all immediately, just a can or two for people who come to the house for it. The rest remains until more is needed and by that time it may have crystallized to some extent. What is the answer?

A. There is no satisfactory answer to your query, except to suggest that health food stores might feature comb honey if they really believe or have evidence that commercially processed honey is damaged to some degree by processing.

"Raw honey" as removed from the colonies is to some degree a perishable product. It will contain wild yeast cells that will cause some degree of fermentation over a period of time unless these are destroyed by the application of heat.

Most commercially processed liquid honey is heated and occasionally to the extent that its original delicate flavor and color are damaged. A temperature of 70° F. for a year or more may do as much damage as 160° F. for a few minutes. Most commercial packers heat honey to kill the yeast and allow the honey to cool rapidly in the retail-packages.

In the handling of a small volume of honey where the basic requirement is to enable proper straining, a temperature of 115° to 120° F. should be adequate. The viscosity decreases very rapidly up to 115° and slowly for temperatures above that. If the honey is granulated, the temperature should be kept lower until all granules have been melted, since a high temperature will discolor the small volume of liquid honey.

HONEY COUGH SYRUP

Q. I would like to know if you have an old recipe on how to make cough syrup with honey?

A. The following directions are taken from the book, Folk Medicine, by Dr. D. C. Jarvis. Boil one lemon slowly for ten minutes. This softens the lemon so that more juice will be gotten out of it, and also softens the rind. Cut the lemon in two and extract the juice with a lemon squeezer. Put the juice into an ordinary drinking glass. Add two tablespoonfuls of glycerine; in terms of drugstore measuring, two tablespoonfuls equal one ounce. Stir the glycerine and lemon juice well, then fill up the drinking glass with honey.

The dose of this cough syrup is regulated according to your needs. If you have a coughing spell during the day, take one teaspoonful. Stir with a spoon before taking. If you are apt to be awakened in the night by coughing, take one teaspoonful at bedtime and again during the night. If your cough is severe, take one teaspoonful on rising, one the middle of the forenoon, one after your midday meal, again in midafternoon, after your supper and at bedtime. As the cough gets better, lessen the number of times you take it. I have observed several points which make it the best cough syrup I know of. It does not upset the stomach, as many cough syrups do. It can be taken by children as well as adults.

FERMENTED HONEY

Q. Would you please advise me what to do with late fall honey? What I mean is honey that was gathered too late to be capped before the cold weather. There was about sixty pounds, uncapped, but it would not spill from the combs when they were shaken, and most of it fermented. If the answer is to extract it and then dispose of it, my other question is what to do with a larger quantity of extracted honey?

A. The honey that you have is fermented and if extracted, can be heated to 145° F. for 30 minutes. This should kill the yeast so that further fermentation will not take place. The honey can then be fed back to the bees next spring, making sure they have opportunity to make cleansing flights. The fermented honey will give them dysentery but this will have no appreciable effect on them as long as they can freely fly.

For honey that is fermented and in the comb, you could follow roughly the same procedure, except for the heating process. If possible, hold the fermented honey in the combs at less than 50° until you are able to feed it back to the bees. The temperature below 50° will prevent further fermentation. You might also place the combs in an area with a de-humidifier to remove some of the excess moisture. The honey could then be fed to the colonies in the spring by placing the combs on the bottom board beneath the brood nests.

HONEY FOR EXHIBITION

Q. Would you be so kind as to send me information concerning preparing filtering, etc. of honey for exhibition purposes. The main trouble is in obtaining the honey free from air bubbles in suspension.

A. Honey for exhibition purposes should be heated only hot enough to eliminate crystals and make it possible to strain through a strainer of about 100 mesh. Temperatures of 110-130° F. should be sufficient.

Do not permit the honey to drop any great height. Place a sloping board in the tank so that the honey runs, not falls, after straining. Allow the honey to settle for at least 24 hours and bottle it cold.

Any time honey passes through air it picks up air which causes bubbles. A lot of bubbles can be prevented during the extracting process if the honey is warm and the extracting is done in a warm room where the temperature is 75-80° F.

STORING COMB HONEY

Q. I have a concrete porch floor. Do you think it would do to store my comb honey in supers out on the porch without any heat for the winter? My honey is all in shallow frames, chunk honey.

A. There are at least two disadvantages to your proposed plan. They are dampness and temperature. Comb honey should be stored in a warm dry place in order to prevent fermentation and crystallization. The ideal temperature is about 70° F.

HEATING HONEY FOR EXTRACTING

Q. I work 45 to 50 colonies of bees and have two 4-frame extractors. They hold 8 full depth or 16 shallow frames. In my last extracting of the season in September or the first week of October, I have some trouble in getting the cold honey to run from the extractors fast enough so I won't have to hang around so long at the end of the day. I have read about electric bulbs with aluminum foil under them being placed underneath the extractors to warm the honey. Do you know if there are electric cables with thermostats that could be placed underneath the extractors that would maintain temperatures of 80 to 90° F.? The honey is run through a heated clarifier and strainer before going into settling tanks.

A. There are many ways to heat the extractors so that honey will run out faster but have you considered heating the honey while it is still in the supers before it is uncapped and extracted? Honey that is so cold it doesn't run out of the extractor should not be extracted. Either store the supers in a warm room (75-80°) for a couple of days or stack them over a light bulb so that the heat will pass up through the frames and warm the honey. Shield the light bulb so that honey will not drip directly on the bulb.

REMOVING THE HONEY CROP

Q. I am an amateur beekeeper. I have had bees now for three years. I have not harvested the honey because I wanted to build up the hives to three stories — the first two for the bees and the third for me. Now I am ready to harvest. When is the time right here in Texas?

I got a swarm from one of my hives and I put an empty box on top of the original so I could feed the bees inside the empty. A week after I hived them they started three or four rows of natural comb from the roof. Now if I take the roof off it will tear up the hive. What should I do? I thought I might put a new story between the top story and bottom story and then harvest the top after a while.

A. While unfamiliar with the proper time to remove the honey in your area of Texas, we can give you some hints so that you will know if your honey is ready or not. The combs of honey should be at least ¾ sealed or capped over with wax. It would be much better if they were completely sealed. This would indicate that the honey was ripe.

If the honey is not sealed or capped over, then you may dip into one or more of the uncapped cells with a toothpick or something similar and get a small amount of honey on the end. Holding this up, the honey should string out. If, instead, the honey drips off like drops of water, then the honey is not ready to be removed.

With regard to the other problem you have concerning the bees which have built the combs in the empty body on top, we would suggest that you remove the empty body and the cover of the hive as one unit. Do not break them apart. When you have them off of the hive, then place them upside down. This will expose the combs in a position so you can reach in and, using a knife or your hive tool, free the combs from the cover of the hive. These can be lifted out one at a time or even a part of a comb at a time. Assuming that there will be bees in this area, we would also suggest that you remove them by using a bee brush. If the honey is ripe, it may be removed to be eaten; if not ripe, it can be placed by the hive entrance so the bees can carry it into the hive.

The new body which you wish to add can then be placed on the colony of bees and the cover put back on to the new body as you normally would.

HONEY FERMENTATION

Q. For the past fifteen years or so I have operated a sideline honey business which today amounts to 51 stands of bees in three widely separated yards. I have had very little trouble with granulation and none with fermentation until this past year when both have become a serious problem.

In the past month or so I have had to pick up several cases of honey that I had sold to stores because it had apparently fermented. Each jar had a heavy layer of foam on top and the honey had forced its way out through the tops of the jars and down the sides. Upon tasting the honey I discovered that it had a very sour taste as though vinegar had been mixed with it.

A. We were sorry to hear the problems you have been having with granulated fermented honey. Fermentation of honey is caused by the action of sugar tolerant yeasts upon levulose and dextrose, resulting in the formation of alcohol and carbon dioxide. The alcohol in the presence of oxygen then may be broken down into acetic acid and water. As a result, honey that has fermented may have a sour taste such as that which you tasted.

Ordinary yeasts do not cause fermentation of honey because they cannot grow in the higher sugar concentration. Spoilage by bacteria is not possible because of the high acidity of honey. The primary sources of the sugar tolerant yeasts are the flowers and the soils. It can be assumed that the yeasts which cause the fermentation of honey would be common and present in most if not all honey that is produced.

The chief factors in honey fermentation are yeast and moisture content. Inter-related with these are the storage conditions and presence of granulation. It has been published that honeys with less than 17.1% water will not ferment in a year, no matter what the yeast count may be. If the moisture content lies between 17.1 and 18%, honey with a yeast count of 1000 per gram or less will be safe from fermentation for one year. Honeys with moisture contents between 18.1 and 19% moisture, a count must be only 10 per gram to assure a honey keeping for a year. Above this moisture level, more than one yeast spore per gram means an active danger of fermentation. Granulation of honey always increases the possibility of fermentation because of the appreciable increase in moisture content of the remaining liquid portion.

If honey is heated to 145° F. for 30 minutes, or at higher temperatures for shorter periods of time, it will be safe from fermentation if protected from further yeast contamination. However, the simplest and easiest method for coping with the problem on a small scale would be to be sure that the moisture content is low before bottling the honey.

HONEY GRANULATION

Q. The honey we buy in our stores never sugars. We don't like honey that has sugared or granulated even after it is melted. Please tell me how to prevent the honey from our bees from sugaring or granulating.

A. Most of the honey you see in the store has been crystallized, but it has been liquefied by heating to a high temperature to retard further crystallization, and much of it is filtered or strained very finely.

Melting honey should not have any detrimental effect on it, unless you are heating to a high temperature (over 140°) and holding it at this temperature too long. It is the combination of temperature and time that changes honey. If you are going to heat it to a high temperature, then it should be only for a minute before it is cooled.

Some honeys which are high in dextrose, are almost impossible to hold in the liquid state. We would suggest that you heat, strain, and bottle in clean containers while it is still hot. You should store it at a fairly high temperature, about 75-80°.

Fig. 55. A safe and practical method of liquefying honey stored in bulk containers. The honey runs out of the melting cabinet as soon as it is soft enough to flow, avoiding overheating. (Photo courtesy, Ont. Min. of Agr. and Food)

HEATING HONEY FOR PACKING

Q. Our present system for heating honey prior to packing is too small and too slow. We have been setting 60-lb. cans in hot water. We have two other systems in mind and want to know which is the better and faster. One system is a hot water jacketed tank heated by gas hot plates and containing an agitator. The other system is inserting a half-inch diameter coil, 100 feet long, in a 40 or 50 gallon hot water heater and pumping the honey through the coil. Should the honey be warmed before pumping through this coil?

A. If the honey is granulated in the can, or even partially so, neither system would work. Certainly with the coil method, the honey should be warmed before pumping it through a half-inch coil. If the honey is granulated, what you need is a hot room or cabinet into which the inverted cans are placed until the honey is liquefied. The Department of Agriculture, University of Guelph, Guelph, Ontario, Canada, has a bulletin entitled "Melting Honey For Repacking." It describes a cabinet which will hold 10—60-lb. cans, liquefying the honey in 12 hours. Write to the head of the department, Prof. G. F. Townsend for such a publication. However, if the honey in the 60-lb. cans is warm and liquid, either system you describe would do the job.

HONEY GRANULATED IN THE COMB

Q. I have five shallow supers that somehow were mixed with empty supers last fall after the honeyflow and they are full of honey that has granulated in the comb. Could you tell me how to get the honey out and save it? I have never had this happen before.

A. Our suggestion would be that you put the shallow supers back on the bees for food but this doesn't answer your question. It will be necessary to warm this honey so it is at least partly liquid so you can extract it. One way to heat it would be to put an electric light (25 watts) in an empty super and stack the supers on top. The top should be left at least partially open so that the heat from the light passes up through the supers for about 24 hours. The honey would then be warm enough to extract. Put the supers back on the bees after extracting so that they will clean them out for later use.

HONEY STORAGE

Q. As a backlot beekeeper with only three hives, I have these questions to ask.

1. At what temperature should I store my bulk comb honey? Some say room temperature, and others say put in the refrigerator or deep freeze. Which will keep it in the best of condition? Would the same go for extracted honey?

2. I would like a small amount of extracted honey to put around the chunk honey in jars to make it more attractive. With only three hives, would it be foolish to invest in an extractor for such a small amount as I would need? Isn't there another way to extract on a small scale?

I heard of putting the comb in an oven, heating it till the wax melts, take it out, let cool and the wax harden, then with two small holes cut in the sides of the wax pour out the liquid honey.

A. Storage of bulk comb honey may pose a problem for you. In the first place, some honeys granulate more rapidly than others. If you happen to be in an area of the country that has rapidly granulating honey, then there may be no easy way for you to handle the bulk comb honey at all. You asked about putting the bulk comb honey in a freezer. This will, indeed, work and prevent granulation. However, it certainly limits the amount of bulk comb honey that you can store. We would consider it best to say that the temperature you should use would probably be normal comfortable room temperature, around 72 to 75 degrees. Even more important is to keep the humidity of the room very low. Bulk comb honey has a tendency to pick up moisture from the air. When this occurs, the moisture content of the honey gets too high and it will then start to leak from the combs. From the standpoint of prevention of crystallization, pretty much the same thing is true for extracted honey. Freezing it will prevent any crystallization. The best way to handle it would probably be to heat it and then quickly cool it down to a cooler temperature. You will destroy not only any potential yeast germs that would cause fermentation, but you would also destroy any crystals.

There are only two ways available for you to get extracted honey to pour around your chunk honey in jars. One way would be to squeeze it out of the combs but in so doing you sacrifice the combs themselves. An extractor would probably not be an excessive expense when you consider the fact that you could use the extractor for 20 years or more. One of the smaller extractors will last you for at least 20 years and is certainly not much of an investment even for a 3-colony beekeeper.

By all means don't try putting your combs in an oven and heating it until the wax melts. In doing this you will cause the honey to be considerably overheated. This will not only darken the honey, but cause it to lose flavor and aroma.

HONEY STORAGE

Q. Last October I took some fall honey which had granulated and melted it and put it in plastic freezer boxes and stored them in the deep freeze. In December I took out a quart and brought it to room temperature by simply setting it on the table all day. I used about half and let the rest stand. It granulated again in about the same time that it took to granulate when fresh.

From time to time I took out more. The honey remained fresh and ungranulated (at the time I took it out.) Two quarts are still in the deep freeze and show no signs of granulation. A friend of mine has frozen comb honey with the same result. I believe that it is possible that temperatures as low as 20° below zero will actually prevent granulation and keep honey fresh. This is, of course, of no commercial importance, but it might be interesting to people who would be interested in keeping honey from granulating.

A. The late Dr. Braun of Ottawa reported that the deep freeze did prevent granulation. The longer it was stored below zero the longer the honey remained liquid after being removed. It would work well with comb, chunk or cut comb honey. The problem of granulation is not great with the backyard beekeeper. He should bottle his honey and if it crystallizes, liquefy it in a hot water bath as needed.

HONEY IN MEDICAL USAGE

Q. Recently I purchased a novel titled, "Honey For Your Health." I have read it quite thoroughly and was amazed at the facts concerning honey. The most interesting fact that still puzzles me though, is the healing power of honey. Just a few days ago, I had acquired a mild form of blood poisoning, due to a sharp meat cutter blade. I went for medical aid, and was given a strong penicillin for it. While waiting for my prescription to be filled out, I noticed Cecil Tonsley's book concerning honey, and was fascinated. Through curiosity, I put a little liquid honey on my protruding infected wound, and found it be of great relief to the pain and burning sensation. Speaking to my doctor, concerning my infection, he told me certain antibodies and antigens are built up naturally through the system, to fight or prevent other foreign bodies such as blood poisoning or other infections to build up. Continuing in this chain of thought, questions concerning antibodies came to mind. Would it not be possible in this modern day and age, for a honey antibody to be built up naturally, in the system, strong enough to fight off infections or other serious wounds, without the use of potent drugs? While examining this question, I put my own chemical system into priority. I am in good health, have had no serious damage done to myself, I try to eat sensibly and in moderation. Would not then any antibodies be strong enough to prevent blood poisoning? Would the constant use of honey prevent this? Do you think that naturally, a wound such as mine could have gone unnoticed without any damage. I realize you are not in the medical profession, but what I am getting at is for everyone to include honey or a honey antigen, to be built up daily in their system, thereby preventing happenings that could be otherwise painful and costly.

A. You raise an intriguing question in your letter concerning the ability of honey to cause an antigenic buildup in the human body. In the context of preventing infections, etc., this is totally new and we wouldn't begin to feel qualified to pass judgment on it one way or the other.

There has been some work down over the years, concerning the healing qualities of honey on an already broken surface on the skin. This is particularly true of burns, although there was one hospital in London that had been doing some work on other types of injuries as well. Apparently, the honey covers the wound preventing air from getting at it, and in addition has some antibacterial action of its own. One of the modes of antibacterial action is in the high concentration of sugars in the honey. An organism that is placed in the honey would become desiccated, as the honey would draw out moisture from that organism. In addition, it has been demonstrated that honey naturally contains a substance called indamine. This material is reported to destroy bacterial organisms and the like. Likely, the combination of the soothing effects of honey along with the antibacterial action, could explain not only your own case, but other similar type reports.

HONEY — CREAMED OR FINE GRANULATED

Q. I am interested in obtaining information on the method of producing what is known as creamed honey (fine granulated).

A. Essentially, the process is this: "To meet U.S.D.A. standards, honey must have a maximum of 18.6 per cent moisture. If it has a higher moisture content, it is likely to ferment. The speed with which honey crystallizes determines, to some extent, the hardness of the finished product, the faster the crystallization, the softer the honey. Use the Dyce process with 18.5 per cent moisture honey. Heat your honey to 140° F. to 145° F., and cool it quickly to room temperature of about 75° F. Add 5 to 10 per cent starter which is finely crystallized honey and mix thoroughly. Place this in a constant temperature room of about 75° F. The honey should be poured into the container into which it will be sold before placing it in the cool room. The honey should be in a solid state in 2 to 4 days. After it has completely crystallized, the honey may be stored at room temperature."

HONEY EXTRACTING

Q. How can I extract honey without renting or buying expensive extractors?

A. There are two ways in which you can extract the honey without renting or buying expensive extractors. The first is not to rent the extractors, but to have someone who already has the equipment do your extracting for you on a per pound basis. This is quite often done by those who have just a small amount of honey to extract. Another method, quite messy, is to cut the honey from the combs and cut the cappings off so that the honey will drain from the open cells. In the old days this was done and the honey was squeezed from the combs by hand. The honey then was strained through a cheesecloth type of strainer in order to separate the wax particles from the honey.

REMOVING THE HONEY CROP

Q. I would like to know if October is too late to take off honey?

A. October certainly would not be too late to take off your honey. In fact, the bees will probably be clustered down away from the honey, so removing the bees from the honey supers would be no problem for you. We would say that by all means go ahead with it.

HONEY GRANULATION

Q. Why does extracted honey granulate more readily than comb honey?

A. Honey stored in old combs will crystallize more readily than will honey produced on new foundation because the old combs with one or more extractions, even though they have been cleaned out by the bees, will have enough primary crystals to start early crystallization. If the combs are stored wet, the honey along the sides of the cells will form crystals of granulation. This seed will very soon start granulation on the first extracting of the next season. The first extracting of honey will most likely granulate more readily than the second or third extracting.

HONEY GRANULATION

Q. How can you keep comb honey so it will not granulate, and how do you ship it?

A. If you are producing a type of honey that granulates quickly, there is little that can be done with the following exceptions. Comb honey should be

Fig. 56. This drying room with the supers staggered to permit the flow of warm dry air through the supers, removes excess moisture from the honey. The hive bees use the same principle by fanning warm air over the comb surfaces. (Photo courtesy, Ont. Min. of Agr. and Food)

stored in a dry room at a temperature of approximately 74 degrees F. If the moisture content of the honey is more than 18.6 per cent, there are ways to remove moisture from the honeycomb. This removal of moisture will assist in keeping it from granulating and also prevent any fermentation. This is accomplished by stacking the supers crisscross and using a dehumidifier and fan circulated air. The comb honey should then be cleaned and wrapped in cellophane or placed directly in a window carton and shipped in a corrugated shipping container. This will keep damage and breakage at a minimum.

HONEY IN THE COMB

Q. Honey is not honey here (Mississippi) unless it is in the comb. In order to offset the cost of general upkeep, I am more or less forced to produce comb honey. I sell it by the frame or super.

The sourwood flow, the last source of nectar of importance, is over about June 15 or 30. The flow stops with a bang. All the colonies are bursting at the seams with bees at this time. Removing the supers of honey crowds the bees badly, causing them to swarm. Please advise me how to remedy this situation.

A. Put on an extracting super or two after you have removed the comb honey supers, this will at least reduce the congestion of the hive. If there are minor honeyflows after the sourwood, you may get a small amount of surplus extracted honey.

UTILIZING LEFTOVER WINTER STORES

Q. For two consecutive years I have had left in the spring several full combs of capped honey that turn very dark and often the honey has granulated. The bees do not touch this honey even if left all summer. Will you tell me what causes this and what can be done to avoid the situation?

A. Honey left on the hive for a period of time always appears dark, especially if it is in brood frames. It almost always granulates, depending upon the nectar source. Try moving some of the honey into the brood nest. Damaging the cappings by scraping them may also help to start the bees working on the honey. You should consider yourself lucky to have honey on hand. Most beekeepers find themselves short of honey in the spring.

CLOUDY HONEY

Q. After straining honey carefully, I find it still cloudy. What is the cause?

A. If honey that has been clarified and bottled by any method is cloudy, this slightly milky color is due to very minute air bubbles not yet removed by straining or settling. Possibly the honey was too cold or not kept long enough in an open tank before bottling, causing the air bubbles to remain in the honey. This cloudiness can be removed by leaving the caps slightly loose and placing the bottles in the hot sun for a few days. The cloudiness raises to the top and finally disappears. The cap then should be screwed down tightly. This can also be done by placing the bottles in a warm room, near 100° F. or placed near a furnace in the winter time. If later the honey does crystallize, the bottles can be placed in trays of water and heated until the granulation disappears. Be sure to remove the containers from the water immediately after the granulation has disappeared.

Fig. 57. Three stages of granulation in jars of honey.

HONEY CRYSTALLIZED IN THE COMB

Q. Please advise me of the best way to handle honey that has become crystallized in the comb. I have several supers of crystallized honey.

A. Honey that has granulated in the comb is quite a problem. You, of course, can cut it all out and melt it up separating the wax from the honey, but in doing so you completely lose the comb and somewhat darken the honey. If you have much of it, however, this is about the only possibility.

If you have just limited quantities of it, the only other possibility would be to use it as a bee feed during the fall and spring. It should go directly on top of the bottom board with the brood nest then sitting on top of it. The bees will eat it out and haul it up into the upper portions of the hive and store it. This will clean it out entirely and still leave you with a good comb for reuse.

HONEY STORAGE

Q. This is my first year in beekeeping and I am thinking of holding my crop until a later date. I do have a building in which to store honey but it is not heated and I cannot afford to heat it. I wonder if the market quality of honey would be hurt if it froze. The temperature may go to 30 below during winter. Would the market quality of the honey be hurt if it candied? What about leaving the honey in the comb and not extracting it until spring and then market it?

A. Almost all honey will granulate and it does not hurt the quality to liquefy and bottle, as long as you do not overheat it. The honey will not be damaged by cold weather either. However, the idea of leaving it in the combs to be extracted in the spring is not good. The honey will crystallize in the comb and too much of it will remain in the comb when you extract it. It is better to extract, put the honey in cans, let it sugar if it will, and liquefy the honey when it is needed for sale.

WHIPPED HONEY LIQUEFIES

Q. Some of my whipped honey tends to liquefy after two or three weeks at room temperature — some not. Please explain.

A. The two possible explanations for the liquefaction of your whipped honey would be as follows: possibly the granulation process was not extended long enough and at a low enough temperature. Granulation for whipped honey takes about 10 days to two weeks at a temperature of 57 degrees. The other explanation would be that whipped honey would tend to liquefy if the honey's original moisture content was rather high to begin with. If the whipped honey gets thin, store at a cooler temperature — if the honey becomes too hard, store at a warmer temperature.

BULK COMB HONEY FOUNDATION

Q. I intend to produce bulk comb honey in shallow frames. I am wondering if honey produced on starter strips would be of better quality than honey produced on full sheets of thin surplus. In other words, would it have thinner and whiter comb?

A. The natural wax may be a little whiter and thinner but you may have trouble getting the bees to draw the comb straight without foundation. Most producers of bulk comb honey use full sheets of foundation. However, there are some who use foundation which is just short of the bottom bar.

MEAD RECIPE

Q. I have become interested in amateur wine making and would appreciate it if you could give me a recipe for mead (Honey wine).

A. This is just one of many recipes for delicious old-world mead. Meads were possibly man's earliest alcoholic drinks, being known in India thousands of years ago. The history of mead, like that of honey, can be traced through five thousand years. Classic meads may be blended with herbs; many types are recorded, such as sack mead, metheglin, sack metheglin, bochet, cyser, pyment, hippocras, and melomel.

Dissolve one ounce of cream of tartar in five gallons of boiling water; pour the solution off clear upon 20 pounds of fine honey. Boil together, removing the scum as it arises.

If a spicy drink is desired, the flavor is enhanced by the addition of different spices such as mace, cloves and stick cinnamon. These spices, together with two lemons cut in small pieces, are put into a muslin bag and added to the boiling mixture.

Toward the end of the boiling process, add one ounce of fine hops; about ten minutes later put the liquor in a tub to cool. When reduced to the temperature of fresh milk (about 70 or 80° F.) according to the season, add a slice of bread toasted and smeared over with a little yeast.

The liquor should now stand in a warm room, and be stirred occasionally. As soon as it begins to carry a head it should be tunned, and the cask filled up from time to time from the reserve, until the fermentation has nearly subsided. It should now be bunged down leaving a small peg-hole which should be closed also in a few days. In about twelve months the wine will be fit to bottle.

COLOR OF HONEY

Q. Last year my honey was dark and this year it is very dark. I put in new foundation and clean frames each year. Most of the honey turned to sugar in the comb. While uncapping the honey with an electric knife, it resealed itself. I put my extracted honey in jars set in pails of water at 130° F. The honey seemed a little darker but it didn't affect the taste of the honey. The crop was also scant this year.

A. In a poor year, nectar collected from minor honey plants may well produce darker honey. There is also a greater potential of the bees gathering honeydew. One thing that occurs to us is that you may need a new location for your bees. Other than that, there is not much we can tell you about the color of your honey. Honey color is dependent not only on the climate of the particular year but also on the flowers that happen to be available. The same is true for the crystallization of the honey. Some nectars result in honey that is darker and crystallizes more rapidly than others. There is not much you can do to correct this or to control it except move your bees to a new location.

You might number your hives and keep a simple running history of each hive including the color and quality of the honey that the hive produced. During the active beekeeping season write down as many observations as possible and make good use of your winter lull in activity by studying over the history of each hive. For example, if a hive produced a scant crop of dark-colored honey two years in a row, move the hive out of that location to new bee pasture and see if the situation improves.

STORING HONEY IN THE SUPER

Q. Last April I started with two packages of bees; as the season progressed I added another deep super, three in all. In September we had a heavy flow of sagebrush. The bees filled all three supers on each hive full and overflowing as I was away for three weeks. My question is this — this spring the bees didn't use enough honey out of the three supers for winter food. I had to take one super from each hive into the house for storage as I don't have an extractor. I am at a loss how to harvest the honey left over from winter. Can I keep this honey stored all summer and use it for feed this winter?

A. Yes, you can save such honey from one year to the next and there are drugs available from beekeeping supply houses that you can use to treat the stored honey supers with from time to time to prevent wax moth damage. Actually, your best storage might be to leave the honey on the bees and let them take care of it. It sounds as though your bees are producing very well and I would suggest that you borrow or purchase a small extractor to eliminate the problem in the future.

HONEYFLOW

SIGNS OF A HONEYFLOW

Q. How do you determine when a nectar flow is on?

A. If you notice bees coming and going rapidly when they leave the hive, and they do not tarry on the landing board but take off as soon as they emerge, then you can feel assured that there is some nectar coming in. If there is no nectar coming in, the bees will loaf on the landing board for some time and do not appear to be in a hurry to leave the hive.

PREDICTING THE HONEYFLOW

Q. When can I expect the honeyflow in my area (Illinois) and when should the honey supers be taken off the hives?

A. I would assume that the flow in your area would not be much different than it is here at Hamilton. Your floral sources are probably very similar to those of our area as well. White Dutch clover is an early yielding nectar source and contributes to good honeyflows from late May on into June. White Dutch overlaps the flow from yellow sweet clover in early June. Likewise yellow sweet overlaps white sweet clover a couple of weeks later. White sweet clover continues into July.

It is advisable to remove your supers of white honey from the clovers by early August to prevent it getting mixed with the darker honey from fall honey plants such as Spanish needle and smartweed. These fall plants give the bees some honeyflow but normally not in quantities worth taking off. In addition, these later fall honeys are darker in color and stronger in flavor and are best used for winter stores.

Fig. 58. A good stand of white sweet clover.

ESTABLISHING COLONIES FROM PACKAGE BEES

Q. I started new colonies from package bees in mid-May. They are in double hive bodies with the supers on top but they did not produce very much honey the first year. Is this normal? Also would it be wiser, considering prices, to rob the bees of everything and start all over with new bees next year?

A. It is quite normal for a colony of bees established from packages not to produce an excessive crop the first year. We normally consider ourselves fortunate if they produce enough for winter stores. With that in mind, it might be pennywise and pound foolish to rob all the honey from the bees and kill them off to try to start again next year. Your chances of getting a significantly better honeyflow next year with an established colony would probably be the better route to follow. You might, however, be able to establish colonies earlier than mid-May and by feeding generously at the beginning, develop strong colonies from packages the first year.

UNCAPPED HONEY FROM POOR FLOWS

Q. What can be done to make bees cap honey filled supers? Also, the queen has started to lay eggs in the supers. What can be done? I have double-bodied hives with supers on top.

A. This sounds typical of a comparatively poor flow plus putting on too many supers. Many times, with a poor flow, the cells are not well-filled nor does the poor flow stimulate the bees' wax glands to produce an abundance of wax. Therefore, the honey cells are left open.

The fact that the queen is moving up to lay eggs in the supers is another indication of a poor flow. In a good flow the bees store the honey over the brood nest so that there are no cells for the queen to lay in. Now, as the honey in the lower part of the supers is used for brood rearing, the queen moves into the empty cell area and deposits eggs. I would guess that the bees are using more honey than they are bringing in as nectar.

The best thing to do would be to remove all empty combs from the supers. Then, concentrate the fullest combs of honey in one or more supers. The combs with small amounts of honey can be placed on top of the inner cover (after it has been placed on the top super of good honey combs). The hole in the inner cover should be open so the bees can go up into the supers with small amounts of honey in the combs. They will bring this honey down and help to fill the empty cells in the supers and brood area.

PROLONGED HONEYFLOW

Q. We have had such an excellent, prolonged honeyflow in this area, (Maryland) with too few supers, that the bees have packed the brood nests with too much honey. Would it be wise to extract some of these combs now or wait until spring and then empty some if the queen needs room? I extracted supers twice this year but could not keep up with the bees. I am a sideline operator with 40 colonies.

A. It may be necessary, if the brood nest is completely filled, to pull out a frame or two in the center so the queen will have a space to lay eggs. This should be done only on rare occasions. If the colony were supered properly, this should not have happened.

It sounds as though you need more supers. We would recommend two hive bodies plus three or four supers to store surplus honey for each colony. You should then extract at least twice — once after the spring honeyflow, and again in the fall. Always leave enough honey to keep the colony supplied with food.

HONEY PLANTS

BEE-BEE TREES

Q. I planted a lot of Bee-Bee trees and the first one at home is in full bloom with clusters all over it. The bees have worked it since the 5th of July. Can you tell me if the seeds will be on the bloom clusters or where? How would you handle them for planting in the spring? I would like to start a project with the Boy Scouts and plant all of the worst land in this area. It looks like an exceptionally good honey plant that would bloom when it is most needed.

A. Bee-Bee tree is the common name given to the Evodis Daniellii and it is reported to be an excellent honey plant. The seed should appear where the bloom cluster was but it is our understanding that the seeds do not stand storage well. It might be better to plant the seeds this fall in a well-prepared seed bed and let nature take care of them. Additional help in such a replanting project might be obtained from Fred W. Schwoebel, Langstroth Bee Garden, The Morris Arboretum, Philadelphia, Pennsylvania.

Fig. 59. The bloom of the Bee Bee tree at the Morris Arboretum. (U.S. Forest Service photo)

HONEY LOCUST — BLACK LOCUST

Q. I have ordered some black locust seedlings to plant near the bees. Do you know whether the honey locust yields nectar and whether it is preferable to the black locust? I believe that I read sometime ago that honey locust was a misnomer and really yields no nectar at all.

A. You were right in your memory concerning honey locust. This is a misnomer. Black locust, in favorable spring seasons, yields quite heavily. However, the honey locust does not yield nectar at all, at least, not in our locality.

HONEY PLANTS

Q. I want to plant some honey plants and trees around my apiary and would like to have your recommendations on what plants and trees would be best.

A. The following plants fill in with nectar and pollen for the bees at a time when the major honey plants are not in bloom:

The pussywillows are excellent for providing pollen and nectar very early in the spring. The French variety has larger flowers than do the common varieties. If your beeyard is near a moist area or a stream, purple loosestrife does well under these conditions. The plants start blooming here in late June and the bees work it continuously until October or when frost kills it. It is a hardy perennial and will easily reseed itself. The plant produces numerous spikes on which the flora starts to open at the bottom and continue up as the flora opens.

Chivirico is another plant which produces flora on numerous spikes. The bees work it readily from the time it starts to bloom in August on until late fall.

Vitex is a small shrub which kills back each winter here, but the new growth in the spring produces many tiny blue lavender blossoms. This plant is in bloom from mid-summer to the end of fall.

A couple of trees which are worthy of mention are the Russian olive tree which blooms in early June. It has a delightful fragrance and the bees work it readily. It is used quite often for windbreaks as well as for specimen trees. The other is the Bee Bee Tree. This is a medium sized tree which has large clusters of small whitish flowers during the latter part of July and the first part of August.

ARGENTINE RAPE — WHITE DUTCH CLOVER

Q. I am interested in planting Argentine rape and perhaps a clover (white?) to supplement my annual honey crop. How much soil preparation, what time should I plant and how, and must I add fertilizer or chemical to the soil to attain success?

A. Both Argentine rape and white Dutch clover should be sown on land previously disced and harrowed, either drilled or broadcast and then covered by a light harrowing to not cover the seed too deep. The seed of white Dutch clover which is finer or smaller on some soils may need some firming of the seed bed by use of a roller.

The best time for planting white Dutch clover is in early spring. For Argentine rape for bee pasture, we would recommend making successive plantings at 3-week intervals starting in April. Argentine rape may vary considerably in nectar yields in different localities.

Fertilization needed depends entirely on soil where you wish to plant; in most cases it would not be necessary. White Dutch clover is a legume and seed should be inoculated (bacteria culture) unless the nitrogen-fixing organisms are present in the soil where planted.

Fig. 60. White Dutch clover. White clover honey is of the best quality; light in color, mild in flavor, and always commands a good price.

HONEY PLANTS

Q. I would like to know if there is anything I can plant (Tennessee) from which my bees would make honey.

A. It would be difficult to tell you of all the plants that produce nectar but Frank Pellett's excellent book, *American Honey Plants,* is available from Dadant & Sons, Inc. and describes every plant in North America that is known to produce either nectar or pollen. Almost all plants which bloom will attract bees. The plants in Tennessee that would be attractive to bees would probably include: basswood, locust, tulip poplar, persimmon, maple, willows, sourwood, sumac, berries, the legumes (clover, vetch, alfalfa) and many others.

Ordinarily it does not pay to plant crops solely for the production of honey. You must harvest seed, hay, or something else from the plant to make it pay. The beekeeper has always depended upon the forest and crops planted by farmers and fruit growers to provide him with his sources of nectar. The bees repay the farmer well with better crops as the result of better pollination.

HONEY PLANTS

Q. I have at present 7 colonies of bees. I have recently inherited a 70 acre farm in Newago County, Michigan. Most of the land is cleared and has been cultivated. Part of it is second growth oak and maple. I would like to plant some trees and shrubs for permanent bee pasture. I have been considering planting 100 Russian olive, 100 Siberian pea, 100 little leaf linden and 100 honey locust trees. By planting seedlings, how soon would any of these become a honey source?

I would like to expand my apiary as I grow older and learn more about bees. I hope to plant about 3 acres of sweet clover next spring.

I do not expect to ever use this land for regular farming, but rather as a place to retire with my bees. Do you have any suggestions or advice?

A. With regard to the farm land that you wish to plant with bee pasturage, we do not feel that the Russian olive or Siberian pea trees will offer your bees much pasturage. Of course the basswood trees that you mention are an excellent honey plant. However, they cannot be depended upon every year. Also, the black locust trees are desirable. You mention honey locust in your letter, but we would like to point out that honey locust is a misnamed tree and does not yield honey. The black locust tree, on the other hand, grows very rapidly, blooms when it is only two or three years old, and is an excellent source of very early spring nectar.

HONEY PLANTS

Q. I have a round bottom ditch about 150 yards long and 15 feet across. Water stands in the bottom during rainy spells. I'd like to put this land to use. The weeds grow as high as your head in a short time. Should I put honey trees on the shoulder? How long would it take them to bear? Is there a chemical I could use to kill the weeds until I get a new growth of plants started?

A. Two generally good honey plants with which we are familiar, Bee Bee tree, small medium sized tree, and Vitex Negundo incisa, shrub or small tree, should grow in your section though perhaps not in all situations. If the shoulder where they would be planted is several feet above the water level when water stands there, it would seem worth planting a few on a trial basis if you can find a way to keep the weeds out until the plants are big enough to hold their own. Vitex blooms at an early age, usually first year; Bee Bee Tree, after several years.

Also, we would bring to your attention: Purple loosestrife, which in some places spreads wild along waters' edge, in wet soil and may grow to six feet tall. We do not have much information as to how this plant does in the South but it is generally a good bee plant. Chivirico stands up well in competition with many kinds of weeds. Other possibilities: willow trees or shrubs which are generally easy to grow. Willows often yield some nectar and pollen for bees in early spring. Spanish needles grow as weeds in low lands along streams. Some varieties and in some localities are good honey plants. You could try it in your section by collecting seed and scattering it.

Most sprays used in killing weeds would also be detrimental to the other plants. For annual weeds there are herbicides which kill the germination before the weeds come up, if applied before weed growth starts in the spring. These also should be used with some caution lest the material be detrimental to root growth of newly set plants.

Fig. 61. Alfalfa has been an important nectar plant in both northeastern and western states.

HONEY PLANTS

Q. I am a side line beekeeper located in southern Ashtabula County, Ohio. Our ground is very hilly and extremely coarse. Alfalfa grows effortlessly and profusely here in this area. This is one of the few places in this part of the country where we can obtain a small surplus of good alfalfa honey.

In addition to our abundance of fall wildflowers we are looking for a fairly dependable source of flora for summer bee pasturage. I am interested in planting sweet clover for the bees only. I have read much about sweet clovers but am perplexed as to which variety or type I should plant to obtain the longest period of pasturage for the bees, plus the most dependable type as a nectar source.

I can obtain white and yellow biennial and Hubam sweet clovers. Do you suppose this soil would support good stands of sweet clover? It grows well here on the side of railroad tracks and in deserted gravel pits. However, where I spread it on waste ground it never took. I have about thirty-five black colonies on the family farm here, and I wondered if five acres of biennial sweet clover, or two and one-half acres of annual and two and one-half acres of biennial would yield enough nectar to give a good surplus of sweet clover honey for this many colonies. I would like any information as to what and how much to plant and when to plant the sweet clovers.

A. It is a little hard to answer your question about sweet clover from this distance for your particular locale. We can pass on to you some general information that has been collected regarding the sweet clovers. In the first place, the preferred crop would be the white or the yellow biennial type of clover. Perhaps both should be in the picture, since the yellow blooms first and then is followed by the white sweet clover. Of course, this would result in a longer honeyflow spreading over some 8 weeks of time rather than the shorter flow that you would get from planting only one of the clovers.

From what you say about your country, the reason it grows well on the side of railroad tracks and in deserted gravel pits is simply because there is more lime in the soil in those areas. Sweet clover must have a lime soil to do well and, when it does grow on an acid type soil, there is seldom any nectar in the bloom. We would certainly suggest that you contact your local county agent who will be able to give you some valuable information concerning the soils in your particular area. He also has facilities to test the soil and to tell you how much lime must be added per acre in order to "sweeten" the soil. Generally speaking, it takes more lime per acre to produce nectar in sweet clover than is usually recommended by the farm and agricultural experts.

Five acres of the yellow and white sweet clover would supply a nice addition of nectar for your 35 colonies if the soil were properly prepared and sweetened with lime. That seems to be about the best we can do from the standpoint of recommendation, and we are sure that the sweet clovers would be your best plant to use.

HONEY PLANTS

Q. Is Crown Vetch a good honey plant?

A. Although some of the Vetches, especially Hairy Vetch which is grown for cover crop and forage, are apparently good honey plants at least for some localities, we have a plot of Crown Vetch which has grown for about fifteen years. During that time we have not observed enough consistent bee activity on the flowers to regard it as a honey plant. And I do not recall having received any reports indicating Crown Vetch as a honey plant. It grows thick in our plot and looks very good for erosion control use. This is what I can tell you. There may be more to learn regarding the potential of Crown Vetch as nectar plant for some other localities as we have only small plot of it.

VITEX

Q. I have two shrubs, vitex (vitex incisa negundo). They are excellent nectar producing shrubs and bees work them heavily. They produce lots of seed that are small, round and very hard, but I have had trouble getting them to sprout.

A. Melvin Pellett of Pellett Gardens, Atlantic, Iowa stated that he has never failed to get a stand of vitex from seed but sometimes some stands are better than others.

The seed is planted in early spring in well cultivated soil that is as weed free as possible. The seeds are placed 1 to 2 inches in the ground and the seed bed is kept well sprinkled until they sprout and the plants begin to mature. The Pellett Gardens use an irrigation sprinkler system which gives a slow, steady flow of water for moisture purposes. The statement was made that if a heavy watering was done, the soil would crust and this would inhibit the plants sprouting and growing. It is necessary to keep the bed weeded until the plants get a good start. Apparently vitex is not real easy to grow but it is entirely possible.

FALL HONEY PLANTS

Q. This fall as I walked near my colonies, I noticed a very strong odor coming from the hives and it wasn't necessary to open the hives to smell it. Could this be disease?

A. Both AFB and EFB, in heavy infections, do emit odors that can be detected by walking somewhere near the colonies. It would certainly be a good idea to check for disease with a thorough examination of the colonies. We would suspect, however, that the aroma you noticed was coming from newly gathered fall honey. A couple of common fall honey plants in your area (Illinois) are heartsease (smartweed) and Spanish needle. Both yield a dark colored, highly flavored honey. Some people find the honey flavor from these sources objectionable but many people do like it. Bakers frequently prefer the heavily flavored fall honeys since they impart their aroma to the baked goods. The honey from these fall sources makes excellent winter stores for the bees.

SPANISH NEEDLE

Q. One of my colonies in a three-week period of time drew out and capped nine frames. Some of this came from heartsease and some from a plant about four or five feet tall that has bright yellow-petaled flowers something like a daisy. Would this plant be what they refer to in bee journals as an aster? Also, I have some goldenrod in our vicinity (Iowa). Does this ever produce some nectar or a surplus in my location?

A. The bright yellow-petaled flowers are probably Spanish needle (Bidens aristosa). The florets of the fall asters which produce most of the nectar are white. Although goldenrod is quite prolific in eastern Iowa, generally speaking it should be considered a minor source of nectar.

BUCKWHEAT

Q. Last year I planted a patch of buckwheat (Illinois) which bloomed well all season, but the bees did not work on it. I did not see one drop of buckwheat honey. Why?

A. Buckwheat is a plant which does rather poorly in our part of the country. It does best in the states of New York, Pennsylvania, Ohio, Wisconsin, and Minnesota. Buckwheat has the reputation as being a crop to grow on poor ground. Many years ago, though, there was an area near here where the farmers would work up the ground each summer and plant buckwheat. At that time the bees in that area would make a fair amount of honey but it was of rather poor quality and quite dark in color.

HAIRY VETCH OR HUBAM CLOVER

Q. Do you think it would pay if I bought seed (Hairy Vetch), for 30 acres of unused soil that is behind me and let the owner plant the seed and harvest it, and I take whatever the bees make? How many hives could work it and what is the crop average for Hairy Vetch in this part of Texas (Mesquite)? If you think this could be to my advantage but think another crop would be better, I am open for any suggestions. I did not suggest clover because it is relatively expensive compared to vetch. Hubam clover does well here because there is about 3 acres not too far away that does fairly well on nectar production.

A. Hairy vetch is an excellent honey plant down there, but at the same time, this is also true of Hubam clover. May we suggest that you contact Mr. Curtis Meier, Manager at Dadant & Sons, Inc., 1169 Bonham St., P.O. Box 146, Paris, Texas, 75460. In addition to being branch manager, Mr. Meier is a commercial beekeeper in Texas and will be able to give you knowledgeable answers relative to honey plants in the state of Texas.

CLOVER FOR HONEY PRODUCTION

Q. I would like to plant some clover this fall for my bees. Which is best, white sweet clover or Madrid clover? Will the Madrid clover bloom the first year? Which is best for my climate?

A. We do not think that you can afford to plant crops just for honey production. However, Madrid clover is an annual sweet clover and will bloom the first year. White sweet clover is a biennial and will bloom the second year. Either of the clovers is suitable for your climate (Dallas, Texas), but we would suggest that you contact your local County Agricultural Agent or State University to get their reaction to the desirability of these plants.

SNOW-ON-THE-MOUNTAIN

Q. My bees are bringing in honey right now that has a reddish color and has a weed taste and odor. It is edible but not top quality honey. I'm wondering what plant or weed the bees got it from. Right now we have in bloom (Oklahoma): ragweed, snow-on-the-mountain, broomweed (August flower), acres of a silver-gray weed which grows 3 to 4 feet tall with flowers on limbs on top, tamarisk, goldenrod, a yellow aster (looks like a daisy bloom), sunflower (acres of them). There are also watermelons near enough for the bees to get juice from. I don't know of any smartweed they are close to but the honey has a smartweed odor. Also mimosa trees are in bloom. Hope you can give me a clue on this.

A. We went to *American Honey Plants* by Frank C. Pellett for an answer to this one and snow-on-the-mountain produces a strange flavored honey that has a red color. It is the only plant on your list that is reported to produce honey which is red. Tamarisk produces a low grade, dark honey with a minty flavor according to Pellett.

INSECTICIDES

INSECTICIDES

Q. Our tobacco-farming neighbors (Zebulon, N. Carolina) use insecticides. Will they harm our bees?

A. We have no way of knowing about the particular insecticides your tobacco-farming neighbors will be using but, to be honest, yes, there will be some harm and some losses. If spraying is done properly, there should be a minimum amount of loss to the bees and the beekeeper. Sometimes a part of the field force of the colony is lost but this is usually replaced in a short while by the emerging bees within the colony. Try to get your neighbors to cooperate with you by telling you the kinds of insecticides they will be using so you can become familiar with the effects and try to get them to warn you in advance of the times they are going to be spraying. You will then be able to confine your bees for a short time until the major effects of the spraying have evaporated. You might find that a gift of honey now and then to your spraying neighbors will increase their cooperation with you.

GARDEN INSECTICIDES

Q. Is there an insecticide for use on gardens that will not harm bees? Our major problem has been aphids on beans and white worms in the ground that eat radishes, etc.

A. Most of the commercial seed and nursery companies issue catalogs which carry a number of different kinds of spray material to use on gardens. Some of these are listed as suitable for use on garden vegetables for human consumption. Rotenone, for example, is suggested for control of a wide variety of chewing and sucking insects. Rotenone is considered to be relatively nontoxic, and can be used around bees with a minimum of injury.

DEAD BEES IN FRONT OF THE HIVE

Q. Two of my colonies show good strength and good vitality, but there is an enormous amount of dead bees in front of each beehive? Could you tell me the reason?

A. We would have to have more information before we could give you the reason for dead bees in front of a hive. We would suspect spray poisoning such as an insecticide used in areas where the bees were foraging. Check the area for any recent spraying and also send a sample of about 50 bees which are outside the hive but not yet dead to the Agricultural Research Service, Entomology Research Branch, Beltsville, Maryland, for tests.

INSECTICIDES

Q. In the past few years several systemic insecticides have come into use such as Di-Syston and Metasystox.

If pesticides such as these are used on pumpkins, winter squash and ornamental gourds and are in effect while the plants are blooming, what is the effect on honey bees visiting the flowers? Are these insecticides incorporated into the nectar and pollen? And, if so, are they harmful to honey bees?

Also, are there any methods to spray insecticides on cucumbers and summer squash without harming the bees, such as spraying very late in the evening or very early in the morning? What methods are used by large commercial growers of these vegetables?

A. Of the two pesticides you refer to, Metasystox (methyl demeton) is no longer manufactured. Instead of Metasystox they are probably using Metasystox-R (oxydemeton-methyl) — these three chemicals are systemic insecticide-acaricides.

It has been our experience that these will seldom poison the nectar and pollen when used at recommended dosages. None of these systemics will be a serious hazard to bees as contact or residual poisons if they are used only in early morning, late evening or during the night when bees are not present in the field. Also, there will be less hazard with each hour of time elapsing between application and opening of flowers.

In California most commercial applications are made during the night and are sprayed by airplanes. Of course, each pesticides must be registered for use on the particular crop, and, in addition, there may be restrictions on the number of applications permitted and there may be a time interval required before harvest.

EFFECTS OF INSECTICIDES

Q. I have a hive of bees nearly wiped out by insecticides. There are nearly 20 colonies in the yard but to date only one has been noticeably affected. Why? What was in the field spray that only one hive was affected? A neighboring field was sprayed by a plane but I would have thought that all of the hives or none of the hives would be affected.

A. When we got into the breeding of bees for pollination purposes, we immediately noticed one very apparent fact. Honey bees might be located in the same beeyard but they might forage from entirely different plants in different locations. In California where we were able to follow this foraging behavior, we found that some of the colonies would be foraging, for example, due north into an alfalfa field, while other colonies were flying due west into a safflower field, etc. Where spraying with insecticides is used, it can easily happen that one or two colonies in an apiary would happen to be flying into the insecticide area and thus be hurt by the insecticide while other colonies were foraging in safe areas and escaped damage.

PREVENTING LOSSES FROM POISON SPRAYS

Q. I believe I have devised a plan to prevent bees from suffering a major loss from poison spray. I have one yard of bees about 1¼ miles from a commercial peach orchard where they spray, causing a heavy loss among my bees in past years. This year I fed the bees heavily with thin syrup and gave a cake of pollen before the spraying took place and they suffered very little. I figured the bees mixed the new nectar with the syrup fed and made the poison less potent and at least protected the young bees and brood. Of course the weather was also cool which helped to cut down on bee flights.

A. Your experience pretty well parallels our own. Three or four years ago we had bees down in central Florida that were badly hit by a spray used on sweet corn. We found, as you did, that if we kept pollen on the colonies and a continuous feed of thin syrup, that the effects could be pretty well nullified.

INSECTICIDES

Q. How can you keep bees when every acre around you is sprayed by plane three or four times a year?

A. You certainly have an excellent question concerning keeping bees in an area in which the ground is sprayed by plane three or four times a year. In Illinois, the farmers are supposed to let the beekeepers know when any such spraying will take place in order that the beekeeper may either move the bees out until the spraying is completed or take protective measures. Sooner or later this question of indiscriminate spraying will have to be resolved in order to protect the bees from being completely wiped out. Agriculture cannot allow this to happen.

As I mentioned above, one method is to pick up the colony or colonies of bees and move them from the location to a different area until the spraying is finished. This is dependent upon the farmers notifying you in sufficient time to do this. Another method is to cover every hive with dampened burlap sacks. This should be done the evening before or very early in the morning of the day that the spraying is to take place. These sacks should be kept dampened if at all possible and should cover the entrance so that the bees cannot get out. Another helpful practice is to put on top of the top super an empty super shell and then put the lid on top of this. This gives the bees room to cluster while they are being confined to the hive.

INSECTICIDES

Q. I have three colonies of Midnites in my backyard and I have had the bees there for twenty years. In one of my hives the bees have been dying by the thousands (I think that number is no exaggeration). Earlier they were making a good start with a good queen but the queen disappeared, presumably killed by the same thing which is killing the workers.

I have never had this experience before and I wonder if it might be Nosema. If it should be, can anything be done about it? The other two colonies are normal.

A. From your letter it sounds as though your bees are being killed by insecticides, but only a laboratory test can tell you for sure. The Agricultural Research Service, Bee Research Division, United States Department of Agriculture, Beltsville, Maryland, is equipped to make such tests. About 50 of the bees which are dying or dead, but not dried up, should be shipped to them for test. They can also test for Nosema, but I doubt that this is the cause of your trouble.

INSECTICIDES

Q. I am a small commercial beekeeper and have had no experience with insecticides. At the present time I have 100 hives on S-1 clover and vetch in the Red River bottom in Arkansas. They are within ¾ mile of where cotton is planted and will be poisoned. What are the chances of getting my bees killed, and to what extent? Are there any precautions to take to minimize the loss?

A. It is quite possible that your colonies of bees could encounter the poison as it is applied to the cotton fields in your vicinity. A lot would depend on the relative attractiveness of the cotton blossom as they are being poisoned, as opposed to the clover and vetch which the bees might also be working. My guess would be that they would prefer the clover and vetch to the cotton, so damage to field bees might be minimized. However, the possibility of insecticide drift on to other fields where the bees may be working, and across the yards themselves, must always be taken into consideration. You would know more about the prevailing winds in relation to the cotton fields and your hives. Also, it would make a great difference whether the fields are sprayed from the air. If you can convince the cotton growers to use ground application, and preferably late in the afternoon or early evening hours, your chances of being severely damaged would be minimized.

NECTAR FROM SPRAYED PLANTS

Q. If the bees gather nectar from sprayed plants, what would be the effect on honey stored in the combs?

A. It is questionable whether much of the poisoned nectar would ever reach the surplus honeycombs. Insecticides applied to non-blooming plants are seldom harmful except when the spray or dust drifts into the hive entrance or onto blooming plants in nearby fields. In the case of early fruit bloom, such as apples some bees might be killed directly in the field leaving very few to make it back to the hive with a poisoned load of nectar. As bees must take the nectar into their bodies, they would usually die of any poisoned nectar before they could return to the hive. Since there would probably be a large portion of the field population lost, there would be little if any surplus honey stored and what honey was stored would probably be placed in the brood area to rear replacement brood for the lost field force. In the case of apple bloom, if the orchardist waits to apply the calyx-cup or petal-drop spray as recommended when 75 to 80 per cent of the blossoms have dropped, there will be little killing of honey bees.

INSECTICIDES

Q. A friend of mine has a colony of bees that hasn't capped any of the brood. Some of it is in the advanced pupal stage, some nearly full grown and still alive. Some of the oldest are dead. Some are bluish and some a dirty brown. It doesn't seem to be at all ropy, and I couldn't smell anything bad. I cut out a piece of it about 2" x 3" for the inspector to send in to the State College.

After looking all through my bee books, I called him up, and asked if anyone had done any spraying up there. He said the Power Co. sprayed their right-of-way with brush killer. I suggested that he kill the queen and give them another one. What do you recommend?

A. Our suggestion would be to send a sample of the bees to the Bee Research Laboratory at Beltsville, Maryland. They should be able to tell you whether they have been killed with an insecticide or a disease.

MOVING BEES

REMOVING BEES FROM BUILDINGS

Q. I have numerous friends around town and in the country who have bees in the walls of their homes, farm buildings and shops. They would rather I wouldn't take off the siding and in many cases this would be impossible. I would like to destroy the bees (or have them removed by introducing some irritating substance) without harming the honey in any way. When the bees have been destroyed, I want my bees to remove the honey and store it in their hives. I'm sure there must be a simple way to do this. Sulfur smoke or the like?

A. Sulfur fumes will kill the bees without damaging the honey. You can probably get your bees to start robbing out the honey by smearing honey about the entrance after the fumes have disappeared. Sulfur fumes, since they are the result of burning sulfur, constitute a fire hazard. There really is no simple way and I doubt that the honey received will be worth the effort.

If you can locate the entrance the bees are using and if there is only one entrance into the cavity where the combs are located, it might be simpler to construct a cone-shaped one-way trap where the bees can leave the cavity but not return. A baited hive set up outside the entrance might entice them to set up housekeeping in a conventional hive.

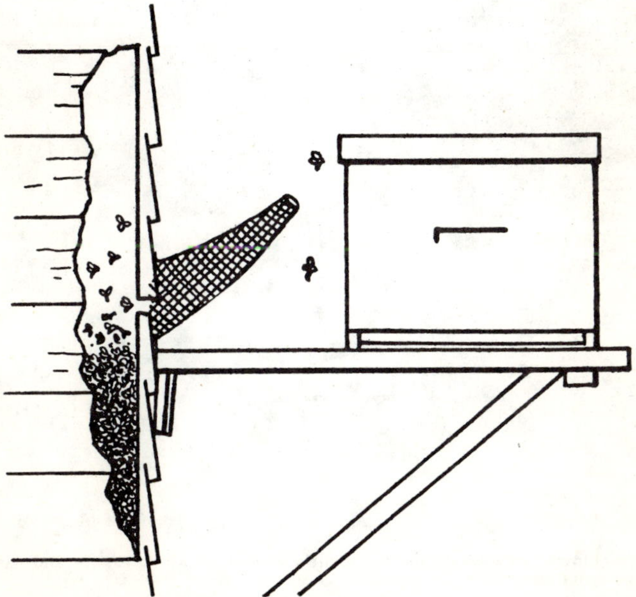

Fig. 62. Diagram of transferring a colony of bees from a house by means of a screen-wire cone. (Drawing courtesy of J. E. Eckert)

REMOVING BEES

Q. I have had fifteen hives of honey bees for the past 10 years. There is only one aspect of beekeeping that I do not like, that is just around the corner.

Getting the bees out of the supers has always been a problem to me.

Using the Porter bee escape, some supers will empty out completely in four or five days. While in three or four hives the bees will not leave the supers.

A. The Porter bee escapes seem to be quite variable in their effectiveness in getting bees out of supers. We do not know if this is due to the differences in the bees, the differences in temperature, or shade and sunlight, or just what. We have seen inner covers having just one bee escape in it from which the bees would leave the supers very rapidly and at the same time have seen other colonies equipped with five bee escapes, one in each corner and one in the middle, and still the bees would be slow in getting out of the super.

It might be a good idea for you to try using Bee-Go (available from bee supply companies.) There is no waiting period. Once the fume pad is placed on top of the super of honey, it is just a matter of a few minutes until the bees have run down into the next super below and the top super is ready to remove.

If you have enough hives to warrant the investment, a bee blower which removes the bees with a stream of air is by far the easiest means of removing bees from honey supers.

Fig. 63. A bee blower in operation. The gentle stream of air is a safe and easy way of removing bees from supers.

MOVING BEES

Q. I tried to move my bees for the first time. What a mess. I put wire in the hole and nailed down the hives. I rented a cart to move the hives to the truck. The nails didn't hold the hives together. The wires came out and let the bees out. I let the bees calm down a little and tried again. I lit the smoker and blew smoke into the hive and tried to put the wire back in, but as soon as I would stop blowing smoke the bees would be out before I could put the wire in.

A. You have certainly had a lot of trouble with the hive that you tried to move which came apart. The most commonly used method of securing the bottom board to the hive is to use hive staples. These staples can be used to fasten the supers to the hive body or supers to supers. They are large enough that they will hold securely against most road bumps.

MOVING BEES

Q. I have enjoyed raising bees for about four years. I think I have done well, but I have made so many mistakes. But now, a new and serious problem has arisen, we also raise horses. I have a beautiful two-year-old colt.

For some reason the bees seem to pick on him and he goes crazy with only one bee. Recently I ended up with a dislocated shoulder and was taken to the hospital.

Also, lately, the bees have been attracted to my hair. Can hair spray affect them?

I love beekeeping and sell lots of good white honey. Can you help me overcome this problem so I will not have to give up beekeeping?

A. Your letter was quite interesting. We think it would be advisable to move your bees to a different location away from the horses. This might be an inconvenience but the risk warrants it. When stung, a horse will not usually run away. Instead, it will kick at the hive. This makes the bees all the more angry and they come pouring out of the hive ready to sting anything in sight. Horses have been seriously injured or even killed by numerous bee stings.

Although we have not recorded cases of bees being attracted to one's hair, it is quite possible. If the bees are angry, they are apt to go for a darker colored object such as one's dark hair. Whether bees are attracted to your hair spray or not is difficult to answer. This could be possible, however.

MOVING BEES

Q. I am in the process of relocating my beeyard approximately 350 feet from its present site because of it being better protected and beneficial to the bees. I have six colonies and my problem is (1) when is it best to relocate the hives to prevent drifting and also a minimum of loss of field bees, meaning at what time of the year and (2) how best to make the move? I am currently clearing the yard and plan seeding it with grass, shrubs, etc. and am using large cement blocks as hive stands and would like to make the move in the easiest manner possible.

A. If they were my bees, I would merely pick them up some evening and move them to the new location. Leave a weak colony behind to pick up the field bees which return. Putting grass or a sloping board over the entrance will help to make the bees realize some changes have been made. Move the last hive after another day or so. You should remove as many of the old landmarks as possible.

Many books suggest that you move them only a short distance each day. You will probably have fewer bees drifting with this system.

MOVING BEES

Q. I wish to move fifteen colonies of bees a distance of seventy-five to one hundred feet. This move is necessary because the owners of the land wish to drill test holes for fire clay where the hives are now located.

I have often moved beeyards a short distance during summer when bees are flying every day. I just simply pick up the three and four story hives with my wheelbarrow and move them about twenty-five feet each day. The bees seem to locate their hives after these short moves without much confusion. I can move a yard of fifteen three or four story colonies this short distance in about thirty minutes. It is necessary to make a number of trips to the yard to make the short moves.

I am wondering what would happen if I made the one hundred foot move at once during cold weather while the bees are not flying. Would they come out of the hives on the first warm day and return to their old location and be lost.

A. When we have moved bees in the past, for short distances, we have done as you have. That is to move in a series of steps. However, I can understand where a fifteen colony beeyard would present problems. You could probably make the move before the weather gets very warm in the spring. However, you can expect that some of the older bees, when they first fly out, will head back to the old location. You might consider putting up a "dummy" hive to catch the stragglers, so they don't become a nuisance for the people in the area.

MOVING BEES

Q. I would like to know about moving bees. This spring I moved a number of hives to a farm about 50 miles from home. I put screen on the front of the hives to keep them in. This was quite a job because some of my bees are hybrids and they boil out with the first touch of their hive. The Caucasians are not like that and I believe I could move them back home this fall. Is there a better way?

A. Most beekeepers use a similar system of moving bees. We cut a screen the length of the opening and about 2-3 inches wide. This is pushed into the entrance in a "V" shape and because it is springy, it stays in place. This can be done before the bees boil out. We also smoke the entrance to the colony before placing the screen in the entrance.

MOVING BEES

Q. Can one move bees for a short distance or a long distance, say 15 miles when the temperature is 30° or colder — say to 10° below zero? If their entrance is screened while moving and after they are located, should this be removed or should one wait until they are settled if they are active?

A. You should have no trouble at all moving bees for the distance that you indicate in this cooler weather. Bees moved at these temperatures are not going to be active. Any that are shaken from the cluster will probably chill and die. I wouldn't think you would need to be concerned about having the entrance screened after they are placed in their new locations.

NUCS

NUCLEUS

Q. I would like to know what a nucleus or nuclei is.

A. The word "nucleus" means a central point about which a gathering or concentration is made and translated to beehives, nucleus means the heart of a colony, a queen and up to five frames of brood. Usually young queens are raised in nuclei. To start nuclei take split-offs or divisions from a regular colony early in the spring. This would mean put three or five frames of brood in a nuc box and give them a young queen. The nucleus will have a good chance at expanding to the point where they can occupy a full colony and the queen in the original full colony will also have a chance to build up the colony again before the honeyflow. So the nucleus is a means of raising young queens and a means of dividing full colonies to make increase.

Fig. 64. A two-queen type of baby nuc box,, with cover raised to show arrangement of combs and feeders in one side. (Photo courtesy of H. York)

Fig. 65. A queen mating yard. Note that the entrances of the nuc boxes face different directions in alternating rows. (Courtesy of H. York)

NUCS

Q. I started with bees three years ago so I haven't had much experience. I haven't had any with nuclei hives. Will you please tell me how to start them? How many bees do I need? Where do I get them (can I use my own bees) and how do I introduce the queens?

A. Yes, you can use your own bees to establish the nuclei. Take two frames of bees and brood and place them in a small hive. Introduce a new queen to the nuc or take the queen from the old colony and introduce a new queen there. Best results are observed if the nucs are made up and moved several miles away to prevent the bees returning to the parent hive. Be sure to feed them until they are established.

We make small hives by dividing a 10 frame hive into 3 sections. Use metal or plywood dividers. The division must be complete so that there is no communication between the different sections or the queens will be lost.

Fig. 66. Observation hive. The best beekeeping education comes through actual contact and a study of the habits of bees. The observation hive may be placed in a window in the spring with an entrance at the front so the bees may forage. It is possible to observe the different changes through which the brood passes, the activity of the queen, and the bringing in and storage of honey and pollen.

OBSERVATION HIVE

OBSERVATION HIVES

Q. At the beginning of the school year I supplied the bees for three observation hives of bees for our elementary schools. They bought young queens. All were accepted and are doing well.

Can these hives be overwintered? Also, can they be kept in the building? I would have to take out some of the bees as young hatch and the rest would undoubtedly have to be fed before Spring. Could they be wintered over a live colony in the beeyard?

A. We want to congratulate you on your efforts to provide observation hives for school children. Not only is this an excellent public relations gesture as far as better appreciation of the beekeeping industry is concerned, but it can also help in creating new honey customers.

We have not attempted to overwinter a small observation hive here ourselves. However, we did so successfully with a larger specially built observation hive. This was one that held six Langstroth deep frames. However, we found that we had to stop up the outer entrance in the cold weather, as the bees would attempt to fly out and would freeze as soon as they got outside where the wind could hit them. They do require additional feeding of honey and pollen, as their consumption will be somewhat increased in the warmer temperatures.

It is also extremely likely that you would have good success wintering them over a live colony in the beeyard. You would probably want to take them out of the observation hive for this. We overwinter small nuclei in this manner using a double screen over the top of a colony. This allows them to make use of the heat from over the colony cluster, yet doesn't get that warm that they are tempted to fly out.

OBSERVATION HIVES

Q. I have been thinking of keeping bees in an observation hive. Please tell me when is the best time to order the hive and bees, and how to manage them.

A. Order the observation hive early enough to assemble it and have it ready before the bees arrive. The best time for the bees to arrive in your area is at the time of early fruit bloom. This would be for a normal package of bees to start a bee colony. However, we suggest that it might be better for your purpose to contact a local beekeeper and see if he can supply you with a frame of sealed worker brood, some attendant bees, and perhaps even a queen to go with it. This would make establishing your observation hive much easier.

One of the big problems with an observation hive colony is that easy manipulation is quite impossible. If the colony becomes too populous, they may try to swarm. If this should happen, watch closely to see if a new queen is reared and properly mated. If one does not show up in the hive and start laying within a couple of weeks, then you will find it desirable to requeen the colony with a laying queen.

Because of the construction of an observation hive, the bees would find it very difficult to form a proper winter cluster. Therefore, it is unlikely that the colony would survive the winter except in a heated building. Since it is likely the observation hive will be placed in a warm room, it can winter satisfactorily if an abundance of stores of honey and pollen are available. The hole in the top of the observation hive permits feeding either sugar syrup or honey.

OBSERVATION HIVES

Q. If I was to make a glass hive, what kind of lighting could I use so it wouldn't disturb the bees. I enjoy bees very much and I read everything I can find on them and I enjoy your magazines very much. Do you know of a book or any thing that would help me on this?

A. The visual spectrum for bees is shortened in the red regions compared to ours, but is extended far into ultraviolet into the spectrum. Therefore, what we see as red would be indistinguishable for the bees. Therefore, your observations could likely be made under red lighting conditions with minimal disturbance of the bees.

Most of the work on the vision and chemical senses of bees has been done by Dr. Karl von Frisch. In 1971, his book entitled "Bees, Their Vision, Chemical Senses and Language" was revised, and is now available for sale.

Fig. 67. The package of bees has arrived, the entrance block and feeder are in place, and the feeder jar is filled with syrup. It's time to open the hive and install the bees.

Fig. 68. The best time to install a package is in the late afternoon or early evening. Make sure all equipment is ready for use and in place before opening the package as installing the package should be a smooth, continuous operation.

Fig. 69. Wet the bees thoroughly with warm water. This can be done with a hand sprayer or you may dip a large paint brush in a bucket of warm water and sprinkle the water on the bees. This wetting prevents the bees from flying.

Fig. 70. Using the end of the hive tool, the next step is to pry up and remove the cover from the top of the package. Be sure to keep the cover near you because there is need for it when the feeder is removed.

Fig. 71. After removing the cover, take the sharp end of your hive tool to lift up and remove the small feed can that was shipped with the package of bees. The remaining syrup in this can may be added to that which you have prepared for the bees.

Fig. 72. When the feed can is removed, place the cover over the opening in the top of the package. This prevents the bees from getting out of the package while you are disposing of the can.

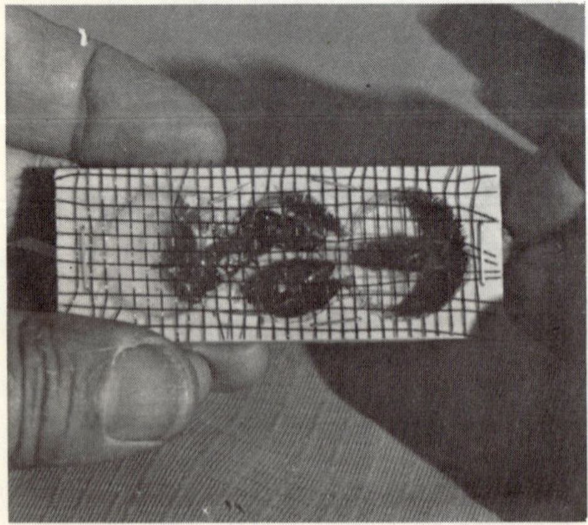

Fig. 73. You will notice that one end of the queen cage has some white candy in it. The cork covering the hole which leads to this candy should be removed and a small hole, about the size of a match, punched through the candy.

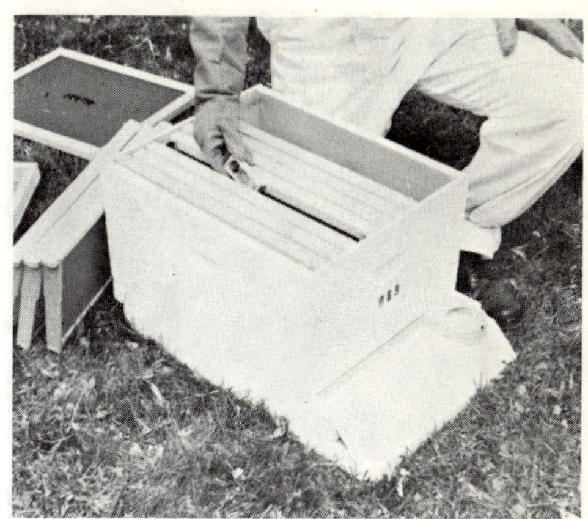

Fig. 74. Now suspend the queen cage in the hive. The proper place for the cage is three or four frames in from the side of the hive. Place the queen cage on a slight angle with the candy end toward the bottom of the hive.

Fig. 75. Take the package cage containing the bees and bounce it on the ground. This will jar the bees to the bottom of the package cage. Pour about half of the bees in the package directly over the frames where the queen is suspended.

Fig. 76. The next step is to jar the bees into the bottom of the package again and pour all of the remaining bees possible over the frames. The bees on the top of the frames will find the queen almost immediately and begin to eat the candy and release her.

Fig. 77. There will still be some bees left in the package. Place the package with the remaining bees in front of the colony and make sure that the hive contains all of its ten frames with foundation.

Fig. 78 Next, place the inner cover on the hive.

Fig. 79. This should be done very gently so that the bees are not crushed. A little smoke on the top of the bees will make them rather quickly run down between the frames and make placement of the inner cover easier.

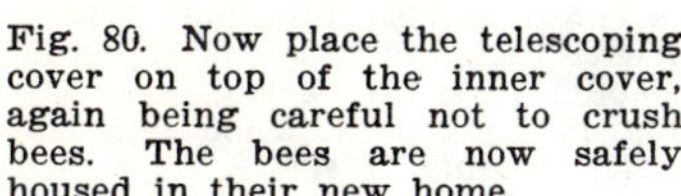
Fig. 80. Now place the telescoping cover on top of the inner cover, again being careful not to crush bees. The bees are now safely housed in their new home.

Fig. 81. Your new colony of bees is entering a period when feeding is crucial until they are established and able to provide nectar and pollen for themselves. Pour the syrup from the feed can into the entrance feeder.

Fig. 82. Your new colony has a lot of work to do. To help them secrete wax and build combs, be sure to use the entrance feeder and to keep it full of sugar syrup for at least the first six weeks.

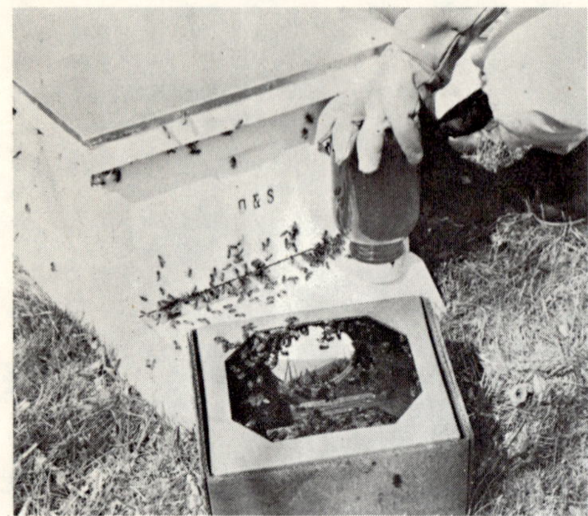

Fig. 83. The entrance feeder jar is carefully placed in the entrance feeder that was supplied with your colony. Note the package cage laying in front with some few bees still clinging inside. They will rather quickly crawl out and join their sisters within the newly established colony.

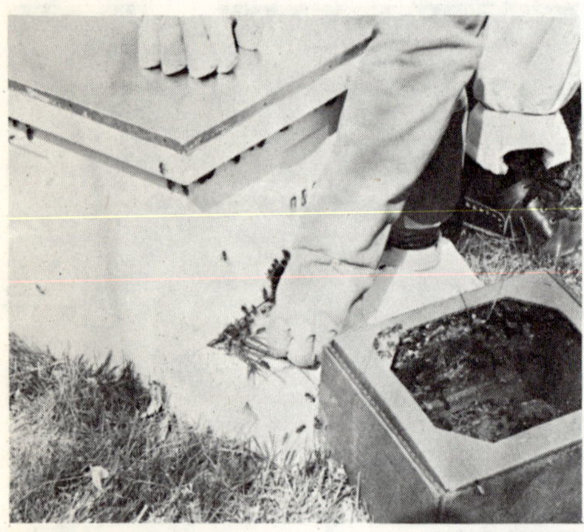

Fig. 84. The last step in hiving your package is to lightly stuff the small entrance with a little green grass. This confines the bees to the hive for a short time and allows them to become accustomed to their new home before they take flight.

Fig. 85. The job is done, your colony is now safely housed and fed in its new home. Success or failure will now depend on the care that you give the developing colony between the time of package hiving and the beginning of the first honeyflow.

PACKAGE BEES

SPRING MANAGEMENT

Q. How many bees are there per pound in package bees?

A. The customary figure used is 3,500 bees per pound. Of course, the number of bees will vary according to the amount of nectar or honey which the bees have in their bodies. Package-bee producers shake bees from combs which have been smoked causing the bees to fill their bodies with honey. Consequently they, as a rule, allow 25 per cent or more overweight so as to be sure of delivering correct weight when the package arrives at its destination.

Fig. 86. A heavily-laden, pollen-covered bee in flight. The sticky grains adhere to the body hairs and are also carried in the pollen baskets on the legs.

ORDERING PACKAGE BEES

Q. When should I order package bees for spring delivery?

A. You should time the delivery of package bees in your area (Kansas) to coincide roughly with the first fruit bloom. This would give the bees a source of pollen which is their protein feed. Of course, you would be supplying them with sugar syrup which gives them the carbohydrate requirements.

If you're ordering only a few packages of bees, the time for ordering is not particularly crucial. However, if very many are desired, the earlier this late fall that you can get your order in, the more assurance you have of getting delivery when you would prefer to have it.

ORDERING PACKAGE BEES

Q. What months should you order bees? Will they live if put in a new hive in July or August?

A. The best time of arrival for package bees would be during early fruit bloom. Later arrivals of packages can still make it, but it becomes increasingly difficult. By July it is obviously too late for you to order and receive the package bees without having to do extensive feeding to get them through the winter. This feeding could be done by feeding them sugar syrup which would be roughly 2 parts of sugar to 1 part of water mixed together. (Heat may be required to get the sugar into solution.) You could take a 5 or 10 pound friction top pail and place your syrup in this. Puncture 2 or 3 small nail holes in the cover and place this feed can upside down over the bees. Place an empty hive body over the feed can to protect it and keep "force feeding" the bees until they no longer take the syrup from the pail.

ORDERING PACKAGE BEES

Q. When is the best time to order package bees in Oklahoma?

A. The arrival of package bees in our part of the country and farther north is usually timed to coincide with early fruit bloom. For us in the midwest that is about mid-April. We would suspect that for Oklahoma it should be somewhat earlier than that. If packages for this coming spring are desired, we would suggest that you get your order entered immediately. The demand for package bees and queens appears to be surpassing the supply and those who don't book their orders early are apt to be disappointed either in total lack of availability next spring or else in receiving their packages at a time that is not convenient for them.

You may be interested to know that many of the northern beekeepers are moving their colonies south in the winter. One of the areas that some of the operators from the Minnesota area are going into is the southeast corner of Oklahoma. They have enough additional time in the spring to divide their colonies and make several nuclei or miniature colonies which are then allowed a chance to develop and build full colonies in time for the flow up North. Once you get your feet on the ground in your beekeeping operation, you may find that you are able to do enough division to replace whatever losses you may be having, or even to enable you to make some increase. However, to do so, it would likely require the feeding of some additional pollen supplement or substitutes.

TIPS ON PACKAGE BEES

Q. I have my package bees hived and think that they are doing rather well. What else should I watch out for or do to help them along to maximum development?

A. About four or five days after installing the package in the hive, you should check to see that the queen is released from her little cage and if she is, remove it. If necessary, feed should be given to them. Incidentally, if it is necessary to hive the package on frames with foundation or use empty combs, feed should be given right at the time the package is hived.

About three or four weeks after installation there will be a critical period, a low spot in bee population. The queen probably will have all the brood area filled that the bees can keep warm. During these first weeks, some of the adult bees that came with the package will have died or been lost. At the same time, none of the brood will have emerged. Various things contribute to this "dwindling" period. Quick changes in temperature can cause field bees to become chilled on their way back to the hive. The normal wear and tear of providing colony needs wears out some of the old original package bees. The more nice, young bees there are in the package, the less serious this "dwindling" problem will be for you.

During the dwindling period, if you have other wintered-over colonies which are strong enough, it will boost the package colony immensely to pull a comb of sealed and emerging brood, shake off the bees, and give it to the package. Put it in the center of the brood area. If the package has dwindled dangerously low, we move the comb of brood with the bees on it. You must be sure the queen from the strong colony is not on the comb. This quick build-up of young bees makes a wonderful improvement in the colony. You'll also need to feed the young colony during this critical dwindling period.

Let's assume we now have six combs in this order: honey, brood, brood, brood, pollen and honey. Colony strength is enough to require more comb for the queen to lay in. Now as your hive contains ten frames, you'll probably find two empty frames on each side of the hive with the honey, brood, pollen, and honey occupied frames in the middle. The queen needs more room to lay eggs in but she will be somewhat reluctant to lay in the outside frames. Splitting the broodnest by inserting a comb between the combs with brood is a poor practice generally. This is especially true if a frame of foundation is being added instead of a comb. This empty comb acts as a barrier between the two halves of the colony. Often the queen is reluctant to cross over and so the other half feels queenless, even to starting supersedure cells. The broodnest can be expanded by moving the empty frames over next to the broodnest. This move puts the honey and pollen on the outside but the bees will still be able to utilize it.

There is also going to be an advantage to having two or more colonies, even if all package colonies are the same age. If you have only one package bee colony, you are more-or-less on a make it or break it level of operation. If you have more than one colony, you will be assured of some sort of crop if one colony fails. If one colony fails, you also have the possibility of boosting its strength with a frame of brood from the other colony, or if all else fails, the two colonies could always be combined into one colony.

Supposing you have a colony that died over the previous winter and you have some of the combs on hand, even though they are a bit moldy. These combs could be added to your package colony, one per hive, and your bees would be able to clean up the combs rather than draw new foundation into comb. A better practice would be, if you have an over-wintered colony on hand, pull a comb of brood to boost the package colony and replace that comb in the over-wintered colony with a frame of foundation or a moldy comb if you have some on hand that should be used up. Since the moldy parts will be torn down and rebuilt, the colony should be fed unless there is a good nectar flow. This is true, also, where adding frames of foundation — there must be an abundance of food (nectar flow or sugar syrup) for the bees to build good combs. Incidentally, the best combs are made when the bodies of frames with foundation are put on a good colony after the honeyflow has started well and before it starts to lessen.

As the package colonies build up, the amount of brood being fed gets larger so the demand for food increases. Anytime the bees run short, their reaction is to cut back their development.

Ideally, the colony should develop from the first egg to reach the point of maximum development shortly after the main honeyflow starts. Excessive dwindling, pollen shortage, lack of room for the queen to lay and room in which to store honey and pollen, or room for the bees to cluster so they don't get too crowded or too hot — thing like these will all work against maximum development.

STARTING PACKAGE BEES

Q. How long should a beginner keep his package bees confined to one single hive body or brood chamber?

A. Until the bees have shown some evidence of drawing comb on all 10 frames, the bees should be confined to a single hive body or brood chamber.

DRAWN COMBS FOR PACKAGE BEES

Q. I have a limited number of drawn out combs and wish to install as many packages of bees as possible about the middle of April. I plan to use four drawn combs in the center of each hive and the other six frames will have new foundation. Will this do or would the bees be apt to build a great many drone cells in the new foundation?

A. Your idea of using the four frames of drawn comb would be of value in establishing the package bees, especially if the combs contained a little pollen. Many packages are hived on entirely new foundation and they get along very nicely. We can see no reason why the bees should raise any more drones than usual, providing, of course, that you feed them well. In our area about 20 pounds of sugar syrup is required to get bees in shape. We continue feeding until all of the frames in the hive body are drawn into comb. There is some danger that the bees will start to store the syrup rather than build more comb but that problem will tend to even itself out as the need for more comb arises.

STRENGTHENING COLONIES WITH PACKAGES

Q. I wish to start fifteen colonies this coming year. My experience is that I get little honey the first year. Would it be advisable, just before the clover flow, to set a two-pound package over the inner cover? If these bees are accepted, the strengthened colony might then collect some surplus.

A. Package bees introduced in early April and fed generously should produce honey the first year. Most beekeepers do not feed enough. A two-pound package introduced just before the honeyflow should certainly be a shot in the arm to the colony.

Fig. 87. The container of sugar syrup has tiny holes punched in the lid. It will be inverted over the center hole of the inner cover and protected by the empty super. Bees will be able to take feed from this container in nearly all types of weather.

FEEDING PACKAGE BEES

Q. When starting package bees on foundation, what is the best way to provide sugar syrup to the bees, and how long should it be provided?

A. The best method of providing feed to a package of bees installed on foundation is to provide sugar syrup in a container inverted over the hole in the inner cover, above the brood nest, and have it available to the bees just as long as they will take it. Usually the bees will stop using syrup whenever nectar becomes plentiful.

An empty hive body is placed around the syrup container to protect it and the hive cover placed on top of the hive body. Many beekeepers use the entrance feeder quite successfully if the temperature is warm enough to allow the bees free movement within the hive. The advantage to the container of sugar syrup placed above the hole in the inner cover is that the bees will be able to take syrup in all temperatures.

FEEDING PACKAGE BEES

Q. I'm just starting beekeeping and have 11 hives. I lost two of them and then got 8 packages of bees in late spring which didn't do any good either. I'm now feeding them with a ½ sugar and ½ water syrup. How can I feed them when it gets cold?

A. What we would recommend for feeding this late, particularly as it gets cooler, is that you immediately switch from feeding half sugar and half water to two parts sugar to one part water. This means the bees have to do less evaporation work to store the syrup away. In order to get a 2 to 1 sugar water mixture completely dissolved, it will require some heat. What works well for us is to heat the water up first close to the boiling point, then remove from the heat and stir in the sugar. Normally this is sufficient. Then, if the weather is quite cold outside, the syrup can be fed in a friction top feeder such as a 5 or 10 pound pail with a couple of small holes punched in the lid. The bees will take the syrup much more readily if it is presented to them while still warm. You could even wrap some thicknesses of newspaper around the feeder can to help hold in the heat. The feeder can is further protected by placing an empty super over it.

PACKAGE BEES

Q. I would like to place an order for a 2½ pound package of bees for April 1 to 15 delivery, but I have had a problem with a swarm of wild bees killing off all my Midnite bees, which are what I have been getting.

Could you please advise whether the Italian bee might be more compatible to the wild bees, or do you feel that as long as the wild bees are there I will have this problem. The wild bees are in a tree in my yard, and I could get rid of them if you think this is the answer.

A. We hope we can offer you some advice that will allow you to properly handle your Midnite hybrid bees and still not get them killed off by that swarm of wild bees. We notice that you have ordered a 2½-lb. package of bees for early April delivery. When those package bees are received, we recommend that you hive them in the evening at dusk. This gives the bees the entire night long in the hive in order to adjust themselves and prepare to defend their home. When hiving that new package, be sure that the entrance to your colony is reduced to an absolute minimum. For this purpose we would recommend that the actual entrance to the colony be reduced down to a size no greater than 1 inch in length and ½ inch in height. All the rest of that entry should be closed off. In this way the small colony in the hive will have to defend only that very small entrance that you have given them. Make sure that no feed is running out of the hive on to the ground in front of the hive and thus attracting any wild bees or swarms that may be around. You should keep the new package with a very much reduced entrance for at least the first four weeks. After that four week period, you can gradually open up that entrance giving them more and more room as they get stronger. Some time towards the end of May, you can then remove the entrance guard entirely and give them the full entrance.

POLLEN

BEES GATHERING COTTONSEED MEAL

Q. We could always use a better honeyflow as our average production is sixty pounds of honey per colony. However, we are never short on pollen as we have many sources (Mississippi) from which the bees gather and store pollen from early spring to fall. I have never examined a colony in early spring and found them short of pollen. Now this brings me to the problem. My bees can get into a room in which I have cottonseed meal and they are hauling it away as fast as possible. They have cut holes in the meal sacks and to see this, one might think this feed room was a Paul Bunyon beehive.

Just what value is cottonseed meal as a brood food? It's my guess they would be gathering this for the same purpose as they would pollen.

I have five hives of the largest grey bees I've ever seen at any time. These bees break cluster at 40° and are among those bringing in the cottonseed meal. Noticing no other bees were flying one morning, I placed a thermometer in the feedroom on the feed sacks among the bees. It registered 38° inside. I believe the breaking of the cluster at a low temperature is the cause of these particular bees using more food during the winter months, as they never had the surplus of food in the spring as will be found among my other hives.

A. You have guessed right, the bees are gathering the cottonseed meal as a substitute for pollen. They seem to gather substitute pollens whether or not they can utilize it. There are, however, beekeepers' reports of bees collecting everything from livestock feed to sawdust and we doubt that they can utilize any of this as feed. There seems to be a difference of opinion on their ability to utilize pollen substitutes; but it is generally accepted today that although bees prefer natural pollen, a substitute is better than starvation. There are few parts of the country that are not deficient in pollen at one time or another. This pollen deficiency is a very serious matter, especially during that period of early spring buildup.

STORING POLLEN SUBSTITUTE

Q. I sent for pollen substitute but they sent me a lot more than I can use for my three hives of bees this year. Can I keep and preserve what I am not able to use up?

A. A good way to protect your extra pollen substitute would be to place it in a plastic bag, tie it securely, and place it in the freezer. Weevils or such insects could not get to it, so whenever you need it again the material should be just as good as when you stored it.

POLLEN FOR FEEDING PURPOSES

Q. Referring to *The Hive and the Honey Bee,* "Using pollen for feeding purposes, use only pollen gathered from colonies free of AFB." Please clarify. Does this mean that, assuming my bees have AFB, the pollen that the bees gather would immediately be affected by AFB? Doesn't AFB inhabit only the inside of the honey stomach or does it also live on the surfaces of the legs or pockets used for gathering pollen?

A. The reason for not using pollen from AFB infected hives is that in the process of packing the pollen on their hind legs, bees incorporate a certain amount of nectar or honey with it. If this honey comes from an AFB infected hive, disease spores may be present. Feeding such contaminated pollen in other colonies could result in spreading the disease.

HONEY BEES AND POLLEN

Q. I have noticed an article has been written concerning the reaction of the honey bee when she loses her load of pollen. What is her reaction?

A. When a honey bee loses her load of pollen (by passing through a pollen trap) she behaves in the hive as though she still had the load — i.e., by going through the motions of dumping the load into a cell. Animal behaviorists would call this a "fixed action pattern."

POLLEN SUBSTITUTES

Q. Please let me know if there is a substitute for pollen, such as soybean flour, cotton seed meal, etc.

A. There are substitutes being used for pollen. Most mixtures include: soy flour (low fat), brewers yeast, and powdered skim milk. The mixture is available from most bee supply dealers. The prepared mixture is best for the small beekeeper, since most of the ingredients are available only in large quantities. Feeding of pollen substitutes and sugar syrup should start in late February or early March in our area.

POLLEN

Q. I have trouble with pollen in my hive frames. I would like to know of some way to prevent this.

A. You could undoubtedly find a lot of beekeepers who would like to trade places with you. One of the biggest problems in much of the country is finding enough pollen for the colonies.

We suspect the trouble with the excess pollen that you described is pollen that is stored up in the honey supers. Pollen storage in the brood nest is highly desirable, and it would be difficult to imagine too much of it down there unless it comes to the point of not allowing room for the queen to lay. By running a two hive body brood nest, this would be expected to help at least somewhat. If you still have an excess of pollen at certain times of the year, and could use it in others, you might pull out some of those pollen laden combs and give them back to the colony in the early spring when pollen isn't available.

One way you can cut down on some of the amount of pollen entering the colony would be to actually trap the pollen before it gets into the colony. The pollen traps that are commercially available are supposed to be something like 70% effective. This means that 30% of the pollen would still get through the trap. If you have as much excess pollen as it sounds, especially during a restricted period of the year, this might be a good solution. The pollen that you trap off could then be fed back to other colonies as it is needed. Storage of the pollen wouldn't be a problem if you have access to a deep freeze. It could just be put in plastic bags and frozen until needed.

One of the techniques that we have used to secure pollen in combs to be used by nuclei and package colonies in the spring is to place an empty super or body beneath the brood nest. This would be the first place that the pollen would be deposited as it comes in from the field, and might be used to prevent the pollen from being carried on up through the brood nest and into the honey supers. This is assuming that you would have some use for the pollen that is stored in the combs. The actual washing of pollen out of the combs would seem to be too time consuming and bothersome to warrant its practice.

POLLEN

Q. What do you think of eating pollen as some health doctors advise? I read from one source that advises to eat two to three tablespoonsful each day and that pollen is 139 times richer in food value than honey. If so, it seems to me that pollen should be used in very small amounts. Or do you believe it is only suitable for bee food?

A. Certainly, pollen is a nutritional substance but it does vary extensively from one source to another, some pollen not even being liked by honey bees. This is why a "blanket" recommendation concerning the use of pollen shouldn't be made. However, a suitable pollen is nutritious, has beneficial properties, and should find more use than as food for bees.

There also is the matter of how pollen is treated. Haydak found that pollen stored one year lost about 75 per cent of its nutritive value for bees, and that pollen stored 2 years or longer was worthless.

More work has been done in Europe on the use of pollen as human food but research in this part of the world has been mainly concerned with pollen as bee food. We are of the opinion that any claims made in advertising for pollen as a health food, or even a nutritional food for humans, would be met with demands of proof by the Food and Drug Administration of this country.

POLLEN SUPPLEMENTS

Q. When feeding pollen supplement, at what specific minimum temperature can a hive be opened without damage to the brood? We will assume that the colony is in a single story hive.

A. It would be best to feed pollen supplement on a sunny day when the temperature was about 45° F. This is the temperature at which bees take cleansing flights.

However, with a single story colony the outer cover and inner cover could be taken off and a cake of pollen supplement put on top of the brood frames quickly. Then the inner cover is reversed with the deep side down and the outer cover replaced as soon as possible so that little heat is lost from the cluster. This could be done in colder weather but I still would suggest a sunny day.

We know a beekeeper in Wisconsin who winters colonies in packing cases. He says he generally starts feeding pollen cakes around the middle of March in his area and feeds about five rounds of 1 to 1½ pound cakes to each colony. It can be pretty cold in Wisconsin at this time of the year and I assume he chooses sunny, moderate days for this work.

POLLEN STORAGE IN HONEY SUPERS

Q. We had quite a rainy season for part of the summer and a neighboring beekeeper told me that bees sometimes store pollen in the honey supers during rainy weather. Every now and then my bees stored a cell or two of pollen in the honey supers above the brood chamber and the honey was fancy grade. I extracted most of this honey and the pollen stayed in the cells which may be a good thing, yet if I had wanted to sell the honey as chunk honey, the pollen would have spoiled it. I did get some chunk honey and by holding the honey and frame under a light bulb could see most of it. Do you know of any better way to detect pollen in cells?

A. You have the answer to finding the pollen cells. A light behind the combs should show all the pollen cells as dark spots. This does not work as well on dark honey but you can pick it out if you look carefully. In the future we would suggest that you produce your chunk honey above a queen excluder. This seems to help keep the pollen from getting into your supers. Also be sure that there are no upper entrances to the hive or the bees will bypass the queen excluder with their pollen loads.

POLLEN STORAGE

Q. Can I store frames containing pollen over winter?

A. Yes, you can store these frames of pollen; furthermore, it's a good idea. Pollen is necessary for brood rearing and is not always available. Store the comb over winter in a cold dry place. Cold weather takes care of wax moth and a dry place will prevent mold. If you must store them in a warm place, fumigate with para-dichlorobenzene to prevent wax moth damage.

POLLEN SUBSTITUTES

Q. What is the difference between animal type brewers' yeast and the yeast used for people?

A. The difference between the animal type of dried brewers' yeast and that used by humans is in the fact that the animal type is bitter and is less expensive than the human type which is debittered and considerably more expensive. The bees do not mind the bitter taste.

The animal type of brewers' yeast may have somewhat less proteins and vitamins than the human type, but for preparation of pollen substitutes it is satisfactory.

POLLEN STORAGE

Q. Will honey bees store pollen in drone comb or foundation having larger cells such as that designated as 706 cells per square decimeter?

A. Honey bees will store pollen in combs built on comb foundation having 706 cells per square decimeter, but none of our bee men can remember seeing pollen stored in drone cells. The late Dr. O. W. Park said that bees in nature store honey in drone cells — not pollen, but we would not be sure that bees might do so under certain unusual circumstances.

POLLINATION

POLLINATION OF RED CLOVER

Q. I am going to produce red clover seed and have been told that I need to get bees with "long tongues." I would also like to know how much increase I can expect from pollination with honey bees.

A. A long-tongued bee is not at all a necessity for red clover pollination. It is true that the nectar of red clover is down a considerable distance in the corolla tube; however, the pollen of red clover is immediately available at the mouth of the corolla and honey bees will work red clover quite well. In general, for effective red clover seed production, you would need about 4 colonies of bees per acre. These bees should be scattered in small groups of 6 to 8 colonies spread throughout the acreage rather than concentrated at one side of the field. If other factors are equal, then the use of the honey bees for red clover pollination should increase the seed yield 8 to 10 times or more above normal.

POLLINATION WITH ALFALFA LEAFCUTTING BEE

Q. We read that alfalfa seed growers have increased their number of the pollinating alfalfa leafcutting bee, *(Megachile rotundata)* thereby reducing the required need of honey bees. Now just what is this leafcutting bee?

A. The Megachile rotundata leafcutting bee is a small bee that is "a cousin" of a honey bee. It uses the alfalfa leaves to chew up and pack to fashion a nest. It then provisions this nest, which would contain one egg, with a little ball of pollen, normally from the alfalfa flower. Because of this, it is a very efficient alfalfa pollinator. The big problem with this, and other wild pollinating bees, is that it is very difficult to control the populations. Man has learned how to propagate them somewhat, but we still aren't able to cope with the natural disasters that take their toll of them including disease. Because of this, the alfalfa seed industry is convinced that the honey bee is the route for the future.

POLLINATION

Q. I am going to plant a vegetable and fruit garden in an enclosed area, which will have small window screens all around it. Can a vegetable garden grow semi-protected like this, or do the bees have to get to it?

A. The planting of some vegetables that you might anticipate in your screened garden will be satisfactory without bee pollination. This would be especially true of the root vegetables. The only reason you would need bee visitation for them would be if you were planning on saving seed from them.

There are other vegetables such as cucumbers and melons which would require bee visitation to carry the pollen from one blossom (male) to other female blooms. Without access by the bees, no fruit would be set. Depending on the thickness of the screen, some of the smaller insects and beetles may be able to do at least a partial job, but bees have assumed the major role in pollination over much of the country.

There are other vegetables that can produce at least a minimum crop without bee visitation but are aided or helped by bee visitation.

POLLINATION

Q. A neighbor wants to pollinate winter grown cucumbers in his greenhouse. We have put in two hives of bees to do the job. Greenhouse temperatures range from about 60 degrees at night to 90 degrees during the day. The ceiling is continuously damp. Our problem is to get the bees to move about freely to work. The bees in one hive refuse to leave the hive; those in the other hive come out but hit the ceiling before they can get oriented. So, all we have are wet wings and bees slithering down to the side of the building.

A. We have had good results with bees in a greenhouse when they had access to the outside as well as inside. We placed the colony outside of the greenhouse with a small opening into the house and the main entrance outside. In this way the main flight of the bees was outside with only a limited number going into the greenhouse. We used this system on tomatoes several years ago. We did use strong colonies since they seemed to give us the best results.

"WASHBOARD MOVEMENT" OF BEES

Q. I got started with bees this spring mostly to pollinate my fruit trees and berries but since then I have found the bees themselves fascinating and derive much pleasure in watching them and observing their progress. One peculiarity in their behavior I have not run across in three or four books I have read is this: The bees group themselves in front of the entrance on the bottom board to one side of the hive. They scratch the board very rapidly with their forelegs and move back and forth about ¼ to ½ inch. Their mouths are also against the board and appear to be working. The hive is new and the weather has been warm so I thought perhaps resin was oozing out and they were gathering it. Close inspection gives no indication that resin is present (pine boards). The board is smooth and dry. They do the same thing to a lesser extent on the painted surface above the entrance. Do you have an explanation?

A. This behavior of the bees is known as the "rocking" or "washboard" movement and many beekeepers have observed in their apiaries, especially in the evening, bees "performing" an interesting activity on the front wall of the hive. The bees stand on the second and third pairs of legs, facing the entrance. Their heads are bent down and the front legs are also bent. The bees "perform" "rocking" or "washboard" movements, thrusting their bodies forward and backward. At the same time they scrape the surface of the hive with short and quick movements of the bent tarsi of the front legs, and their mandibles, with a rapid shearing movement, slide over the surface as if cleaning it. The tips of the antennae touch the surface, moving constantly. After a while one can observe that some material has accumulated on the lower edge of the mandibles. Periodically the bees clean their mandibles as well as the tarsi of the front legs. Occasionally they move and continue this activity at another place. These rocking movements probably serve as a mechanical cleaning process by which the bees scrape and polish the surfaces of the hive. Usually this "scrubbing" dance, both inside and outside of the hive, is "performed" by young bees. (Answer from *The Hive and the Honey Bee,* 1975 ed., Dadant & Sons, Inc.)

PROMOTION

PROMOTION

Q. As an amateur beekeeper and enthusiast I have been asked to present a program for a local civic club. I plan to use an observation hive, however, I would like to present some slides showing basic beekeeping procedures and some frame close-ups showing larvae and the queen bee. Could you tell me where I could borrow or purchase beekeeping slides.

A. Dadant & Sons, Inc. has created eight sets of beekeeping slides on various phases of beekeeping and these are available free of charge. They ask that you do pay return postage, either UPS or insured parcel post and that you return them as quickly as you are finished with them so that the slides can be kept in circulation. It would also be advisable to send in your request as far in advance as possible as the demand for the slides has been considerable. Although more beekeeping slide sets are in preparation, the present list of titles includes:

No. 1 — Package Bees
No. 2 — Supering Package Bees
No. 3 — Spring Management
No. 4 — Supering Overwintered Colonies
No. 5 — Diseases and Enemies
No. 6 — Queens and Requeening
No. 7 — Honey Plants
No. 8 — Starline/Midnite Breeding Program

PROPOLIS

PROPOLIS

Q. Can you please give us any information regarding propolis? Propolis has been reviewed over here (South Levon, England) on television. Also, how could it be consumed beneficially by human beings and for what purpose?

A. Propolis is a generic term for certain sticky substances secreted by plants, mainly trees. It occurs, for instance, on the buds of poplars and on the trunks of pines. This propolis is also used for the material the bees use in the hive, which is usually a mixture of (plant) propolis and beeswax. The word is derived from the Greek pro (before) and polis (the city) — propolis being used to make the protective shields at the entrance to a hive. (Bee Research Association).

Propolis is indeed finding more uses than a sealant, fixative, or nuisance within the beehive. Propolis is being used in both Europe and Asia in minor medical applications. Small cuts or minor burns are reported to respond well to a tincture-administered propolis.

Research on propolis has been done in the Soviet Union for some time. Scientists have used propolis to treat tuberculosis and stomach ulcers. In Scandinavia, much research has been done since the late '60s. Findings from studies conducted in the Soviet Union indicate that propolis may have some effectiveness when taken as a food supplement by individuals who have infections of the mouth, nose, or throat, Infections of the kidney, bladder, and prostate have also reportedly been relieved by propolis. Propolis may well become a benefit to modern medicine. However, to say that propolis is a cure-all or a new and powerful antibiotic would not only be premature but irresponsible. Nevertheless, propolis is attracting increasing interest and research.

It should be pointed out that these reports are based on articles written on the subject of propolis. We have no documented evidence of cures resulting from the use or consumption of propolis and certainly do not encourage home experimentation until further research confirms or disclaims these reports.

ELIMINATION OF PROPOLIS

Q. As the elimination of propolis is desirable in modern beekeeping, I coated 16 hive bodies and frames with hot paraffin except where comb fastens to the frame. This took about 15 minutes per unit and is not expensive. What is your opinion on this?

A. It will be interesting to see how your experiment comes out where you coated hot paraffin on the inside of the hive and also the frame. We would guess that the bees would still continue their propolizing and burr combing and that after a few months such colonies would have just as much propolis and just as much burr combing as their nature and inheritance required in spite of the fact that you had coated the wooden material with the hot paraffin. There will be a residual effect from the paraffin for a short period. This is partly because of the fact that the odor of paraffin is slightly repulsive to honey bees and they will tend to stay away from the paraffin coated wood until such time as the odor has been eliminated. It would be our guess, however, that eventually they will go ahead and use the propolis and the burr combing.

PROPOLIZING THE QUEEN EXCLUDER

Q. Why did the bees close up the openings in the queen excluder that was placed between the brood chamber and the first super? The openings were plugged with a thick wax. This was noticed when I started to put on the second super. There was no activity at all in the first super.

A. One possibility is that the thick wax which you mention the bees have used to close the openings in the queen excluder may be propolis rather than wax. Propolis is a vegetable waxlike, sticky material which the bees use to close down unwanted openings in their hive. Many times they will almost completely close the main entrance to the hive with this material. When it is warm the propolis may be quite sticky. When the weather is cold, the propolis becomes quite hard and brittle.

I would also guess that the honeyflow has not been too great in your area. Sometimes when the honeyflow is rather meager the bees will tend to store all of their surplus honey in the bottom unit and fail to go up into the supers. This is especially true when the frames in the super contain foundation rather than built combs.

We would suggest that you remove the queen excluder and then put the super back on. This will make it easier for the bees to move up into the super. Then, if there is enough of a honeyflow to warrant it, the bees will start to work in the super, building combs and putting the surplus honey in them.

QUEENS

DISCOVERY OF THE QUEEN BEE

Q. When I was small it was common to hear the expression that he or she was the King Bee of the family. Someone told me an Englishman came over here in 1823, ran a research on a hive and found out that the King was laying all the eggs. Since then there has not been a King and up to then there had not been a Queen. Is this correct?

A. In earlier times, the large bee in the hive was assumed to be a King, probably because of our own familiarity with monarchies. However, in Spain in 1586, Luis Mendez de Tures discovered that the queen bee was in fact female and the one that laid the eggs. Apparently the old reference to the King Bee has held on in local vernacular, and the replacement with the queen bee has been slow in coming about. In fact, the only males in the honey bee colony are the drones whose only purpose in life is to mate with the queen should the occasion arise. Otherwise, the drones do no work and the female worker bees wait on them hand and foot until cold weather when the workers drive the drones out of the hive to starve or freeze.

QUEENS FROM EMERGENCE TO EGG LAYING

Q. How soon should a young queen begin laying after she emerges from her cell?

A. A lot of activity takes place between the time the young queen emerges from her cell and she has mated, been accepted as queen in the colony, and has started laying. The only available figures are averages that attempt to take into account: colony conditions, weather conditions, presence of drones, etc.

A virgin queen emerges and immediately seeks to establish her dominance by attempting to destroy all other queen cells in the hive. The first mating flight usually occurs on the 5th or 6th day. In one of his experiments, detailed in *The Hive and the Honey Bee,* Dr. Everett Oertel observed flights of 60 virgin queens and found that virgin queens frequently mated on the first flight. Thirty-two of fifty-four virgins observed, mated within 8 to 9 days after emergence, 16 from 6 to 7 days, and the remainder in 10 to 13 days. On the second to the fourth day after mating, the queen usually starts egg-laying although she may commence laying as early as 14 hours after successful multiple-matings.

The difficulty for the beekeeper is knowing how long to wait for the young queen to begin laying and when to take corrective action. Taking colony and weather conditions into account along with the presence of drones in plentiful supply, let's establish a minimum of eight days and a maximum of three weeks. The exception here would be the queen reared and established in a colony in late fall who may not begin extensive egg laying until the next spring.

VIRGIN QUEENS

Q. I had always believed that the first virgin queen to emerge would kill the other virgins in their cells but I have seen more than one virgin at a time in one of my hives. Does this mean that these sister queens would live side-by-side in the same hive without fighting?

A. There are several possibilities here and whatever conclusion we may reach will at best be a guess, based on previous experiences with virgin queens. We believe that what you saw was a "stand-off," possibly occasioned by whatever disruption there was in the hive being opened. The bees were temporarily thrown off their guard and the situation probably would have developed into a fight had the hive not been opened.

The first virgin queen to emerge may immediately attack and kill other virgins in their cells, or she may wait a few hours, or the hive bees may prevent her from attacking the other virgins by holding them prisoner in their cells. The prisoner virgins would have continued to develop rather rapidly, even though still confined to their cells so that they would have appeared several hours older than what they actually were when they emerged and they would have taken advantage of the disturbance of the hive opening and consequent disruption of the guarding by the hive bees to emerge.

We suspect you were viewing the lull before the storm as a fight for dominance is almost inevitable. There have been several reported cases of mother and daughter queens laying in the same colony but usually only during a heavy honeyflow. When the flow diminishes, there is usually a fight for the leadership of the hive, resulting in one surviving queen.

PROGENY OF PURE-BRED QUEENS

Q. Will my pure-bred Italian queen continue to produce pure-bred Italian sons and daughters after I introduce her into the colony or should I watch for signs of mismating among her offspring?

A. Your pure-bred Italian queen was mated before being shipped to you by the queen breeder and after she once begins to lay, she will not mate again. Her drone sons will remain pure-bred Italians and will have an opportunity to Italianize the apiary and her worker daughters will not mate so they will also remain pure-bred Italians. Her daughter queens allowed free flight could become mismated unless the apiary has become so fully Italianized that the only drones these daughters meet on their mating flights are also Italians.

What it amounts to is that the race of an apiary can be changed by simply introducing only queens of the desired stock. Charles Dadant was offering to Italianize his neighbor's apiary in the late 1800's because he knew it would improve not only their bees but also improve the mating of his own queens. To prevent any chance of mismating and keep only pure-bred stock, it would be necessary to requeen with pure-bred Italian stock rather than allowing supersedure queens free flight.

"BALLING" ACCEPTED QUEENS

Q. Can you explain the term, "balling the queen"?

A. Bees occasionally become hostile toward their own queen and envelop her with their bodies, forming a living ball of bees from one to three inches in diameter about her. Unless the ball can be dispersed with smoke or by dipping the bees in water, the queen will eventually be stung to death or be smothered.

According to Ronald Ribbands in his book, *The Behaviour and Social Life of Honeybees,* balling may occur in colonies which are disturbed: in the early spring; when a young queen is just beginning to lay after a long break in breeding; or when a queen has recently been introduced. Queens have also been attacked this way when returning from a successful mating flight, but not after an unsuccessful mating flight. It is possible to interpret all of these circumstances in terms of response to an unaccustomed odor. This is one of the reasons why much care is taken when introducing a queen and in other operations involving situations where the balling response may occur.

QUEEN BEE FAINTING

Q. Is it true that queens can "faint"?

A. We have heard the term "fainting" and have also heard of "cramps" and apparently both terms relate to the same thing. Researchers don't write about it often but sometimes under stress conditions such as marking a queen or handling her in other ways, she will simply appear to be dead. Probably more than a few "fainted" queens have been destroyed as dead but under the symptom of "fainting," she will sometimes recover some minutes later. Her court of attendant bees hover around, nursing her, and offering food, and she recovers, apparently no worse for the experience. We know of no known cause.

PIPING OF QUEENS

Q. I have read about queens piping before swarming. What kind of noise is it and how far is it audible?

A. The piping of queens is a rather shrill noise and can be heard outside of a closed hive. Many beekeepers have reported hearing it from queens about to emerge from their cells or from those queens that have emerged from their cells and are looking for other queens.

LOSS OF QUEEN DURING SWARMING

Q. I can understand that swarming is Nature's way of dividing colonies to create new ones but why does the colony sometimes end up queenless when the swarming impulse is over? What went wrong?

A. A number of things could have gone wrong as swarming is one of the major disruptions a colony can have. Most likely, the colony was left with a young queen after the swarming impulse had passed and she was probably lost on her mating flight. It would be a good idea to examine such colonies about two weeks after the swarming. Two weeks should give the colony time to settle down, the young queen time to mate and begin laying, and you will still have an opportunity to introduce another queen if the colony has been left queenless.

COLOR DIFFERENCE OF PROGENY

Q. What causes such a great difference in color among the individual bees and also among the colonies in general whose queens are a mother and her daughters?

A. If you have a pure Italian queen, her worker progeny all having the same markings, and from her rear a young queen, and this young queen mates with a pure Italian drone, you may expect to find the same markings in the worker progeny of the young queen as are found in the worker progeny of her mother. But if this young queen mates with a black drone, then you will find the worker progeny different, some of them looking like black workers and some like Italians, and perhaps intermediate markings.

ACCEPTING OF QUEEN AFTER AN ABSENCE

Q. How long could a queen be kept out of a colony before she would have to be introduced in order to put her back into the same colony from which she was taken?

A. We do not know how long the queen could be kept out of a colony from a maximum standpoint. We have frequently kept a queen overnight and introduced her back into the same colony the next day and she would be accepted. This would be without using a normal queen introduction method. In other words, we would just turn her loose the second day. Many times we keep a queen out for a period of three to four hours with almost 100 per cent acceptance when putting her back again by just turning her loose.

REARING QUEENS FROM GOOD COLONIES

Q. I have two colonies in my backyard. No. 1 is by far the most industrious colony of bees I have ever had and the most productive. Is there some practical way to introduce a new queen from this colony to No. 2 colony? Both colonies are Italians.

A. The only practical way to introduce a queen from No. 1 colony to No. 2 colony is to rear a queen from it. This may be done in many ways but one of the simplest would be to kill the queen in No. 2 and place a queen cell or a frame of brood from No. 1 in the colony, making sure that they rear a queen cell on this frame.

MISMATING OF YOUNG QUEENS

Q. How can I prevent mismating of my young queens?

A. The only way to prevent mismating of young queens is to provide drones of the desired ancestry in apiaries surrounding the yard from which your young queens will be taking their mating flights. Even then, there will be some chance of mismating. More and more queen breeders are turning to artificial insemination of queen bees as it allows absolute control of mating. Artificial insemination, however, may be beyond the scope of your queen rearing operation and if so, providing large quantities of the desired drone stock would be your best procedure.

LAYING WORKER

Q. Will bees with laying workers build queen cells?

A. A colony with laying workers will try to make queen cells from their larvae. Normally, these result in elongated cells which produce nothing. Parthenogenesis does occur occasionally in honey bees. However, in these cases, a female is produced.

Fig. 88. The queen and her court. The queen is being fed by one of her workers, examined by others and the bee (top right) is licking queen substance from her body. (Photo, C. G. Butler)

LOCATING THE QUEEN

Q. How do you find the queen?

A. First, it is best to use a minimum of smoke. Too much may cause the bees, including the queen, to start running on the combs, and it is more difficult to focus your eyes on a moving target.

After removing the supers, if any, lift out one of the outside combs — gently, if there are bees on it. Before looking at the bees on the comb you are holding, glance at the near side of the next comb, still in the hive. If the queen is not there, then check the comb you are holding.

In looking over a comb to find the queen, let your eyes scan across the top third from one side to the other; then drop back down to the bottom third and scan back across; lastly go back over the center third of the comb.

If the queen is not found, turn the comb over and check the other side. If still not found, set this comb in an extra hive body or lean it against the hive. Repeat the whole process, comb by comb: remove, check side of next comb in hive, then both sides of the comb just removed.

Do not spend too much time looking at one comb. The longer the hive is open, the more smoke will be needed, and, in turn, the bees are more apt to start running. (At the worst they may leave the combs and move helter-skelter around the inside walls of the hive.)

Part of the success in seeing the queen is in learning to look for the distinctive queen shape among the mass of bees rather than a bee-by-bee inspection. Another help is to watch for the quite common arrangement of attendant bees around the queen.

If none of the above works and you must find the queen, place a queen excluder across the entrance to the hive and shake all of the bees off the combs so that they must pass through the excluder to reenter the hive. When all of the bees have gone through the excluder, the queen and the drones will be found in front of the hive. If you do encounter a great deal of difficulty in locating your queen, have the next queen marked with a drop of paint on the back of her thorax. Most queen breeders will include clipping and marking of the queen for a small fee.

QUEEN REARING

Q. I am keeping bees for a sideline (Florida) and would gradually like to work it into a full-time business. I would like to learn how to raise my own queens for requeening and for making new colonies of bees. I would appreciate it if you could send me some information on raising queen bees as I do not know anything about raising them.

A. We would suggest that you visit one of the many queen breeders in Florida and also read one of the many good books on queen rearing such as *Queen Rearing* by Laidlaw and Eckert, available from Dadant & Sons, Inc.

Most queen breeders graft very young larvae (12-18 hours old) into prepared queen cups. These cups are placed in strong queenless colonies 10-20 to a colony which feeds and rears the queens. In ten days the cells are removed and each

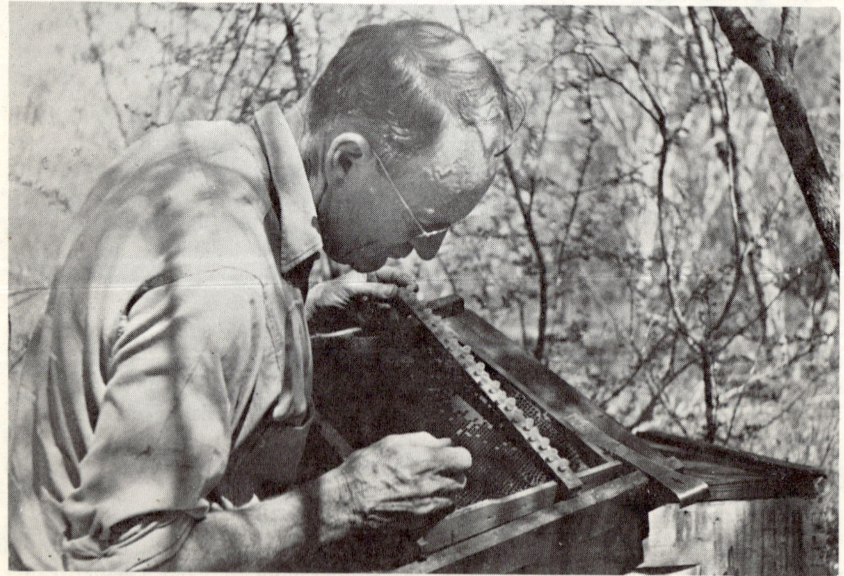

Fig. 89. Grafting larvae into prepared queen cell cups. (Photo courtesy, Chas. S. Engle)

one is placed in a small queenless colony to emerge, mate and start laying. The laying queens are removed as needed and replaced with more queen cells. You can do queen rearing on a small scale by merely removing the queen from a desirable colony and letting the bees rear their own queens. These cells may then be removed in 10 days (carefully) and placed in queenless colonies as above.

Fig. 90. The grafted larvae have been placed in strong queenless colonies which fed and reared the queens. The cells will be separated and placed in small queenless colonies to emerge, mate and start laying.

LAYING WORKERS

Q. One of my hives was queenless for several weeks so I ordered a queen and introduced her to the colony. About a week after introducing her I looked into the hive and found some drone brood but no worker brood. Then a week later I found some scattered drone brood and some solid worker brood. I am sure the drone brood did not come from laying workers. Can you tell me why the queen first laid drone eggs and then, all of a sudden, started laying worker eggs?

A. We would suggest that your colony did, indeed, have laying workers. You state that the colony was queenless for several weeks and about a week after giving them a new queen, you found drone brood and no worker brood. The fact that you found drone brood a week later certainly indicates that eggs were laid before you gave them the new queen. Drone brood is in the sealed state for about 14 days and the presence of sealed drone brood a week after you gave the new queen could not possibly be from eggs of that queen. Another point to remember is that the worker brood now in your colony is close and compact while the drone brood was scattered around the brood nest. This is also an indication of laying workers. You don't have to worry about the problem as it seems to be clearing itself up. If the queen in your colony is now a good one, she will quickly take over and, as the laying workers die, your colony should build up rapidly and well.

MIDNITE HYBRIDS

Q. I have four hives of bees, not the Italian and not the Midnite. I am wondering if I should get a swarm of the Midnite hybrid, and should the hive be put near or some distance from the ones I have now?

A. You should have no problems at all in moving a stand of Midnite bees in with the other bees. In fact, the Midnite queens can even be introduced in colonies of other bees, and as the progeny emerges, will take over the colony completely. The only time some difificulty might be involved is when the queen is superseded and a new queen goes out to mate. Therefore, you may find it to your advantage to requeen your Midnite colonies on a regular basis.

ADVANTAGES OF 2-QUEEN COLONY MANAGEMENT

Q. What are the advantages of 2-queen colony management?

A. According to Robert Banker in *"The Hive and the Honey Bee,"* a 2-queen system of management is designed to obtain maximum honey yield from each hive unit.

"Work done by Farrar demonstrates that as colony population increases, efficiency of honey production increases also. In other words, one large colony will produce more honey than two smaller colonies, each one half the size of the larger unit. Thus, the aim of 2-queen management is to develop colonies with large populations throughout the honeyflow."

According to Banker, the 2-queen systems are generally thought of as being used in small to medium sized production units, mainly due to the extra labor involved. He cautions, however, that such a system may not be practical or profitable in all areas. It is recommended that the beekeeper who wants to investigate the 2-queen management system should start out on a limited basis to find out whether or not the system is good for his or her particular area.

There are disadvantages and advantages of the system. For example, one disadvantage is that the tall colonies which result, are more difficult to keep from tipping over, and level hivestands are essential according to Banker. However, one big advantage is that optimum colony populations are obtained for maximum honey production with swarming and supersedure loss virtually eliminated.

QUEENLESS COLONIES

Q. How can you tell when the colony is queenless without going through the colony looking for the queen?

A. One very quick and easy way to determine if a colony is queenless is to pull out some combs from the center of the brood nest. If there are no eggs or young larvae, you can feel relatively certain that either your colony is queenless, or else it has a failing queen.

GYNANDROMORPHS

Q. I just got through looking through my three hives and all my queens are showing good brood patterns One, however, has young bees which are colored very strangely. Their colors are yellow on over half of the abdomen and black on the other. My question is, is this queen all right or should I requeen? She has a good egg laying capacity so far.

A. The condition which you describe concerning the colors of the worker bees is caused by a genetic condition and the bees produced with this condition are called Gynandromorphs. Besides the color variations which you have described, there are often such things as the head of a drone on a worker bee or the head of a worker bee on a drone or the eyes of one may be on the other type. Normally this does not involve too many bees in a hive. However, should there be a great number of them, it would be better to requeen the colony since there is little likelihood that this will ever change with your present queen.

RAISING QUEENS FROM GOOD COLONIES

Q. One of my colonies produced more than 200 lbs. of honey last year and is repeating the same feat this year. Would it be possible to reproduce queens from this stock in order to requeen all my other colonies? Would that be an intricate process?

A. If you have a colony which produced over 200 pounds for the past two years it might make a good colony from which queens may be reared. Queen rearing is an exacting procedure; timing is very important. You could produce a few queens by placing a super containing plenty of bees, brood of all stages and honey along with some eggs above a double screen on a strong colony. Have an entrance to the rear. The bees will produce queen cells which should be removed carefully in 10 days. Place the cells in colonies which you need to requeen.

BEES CARRYING EGGS

Q. Has anyone ever had a queenless colony of bees go to a queenright colony, take a fertilized egg from them, bring it back to their queenless hive and raise themselves a queen? I think I had this happen last year.

A. The question of whether bees carry eggs from one place to another is one of the oldest controversies in beekeeping. We would have to say that they do not carry eggs from one place to another for the purposes of raising a queen because no evidence has ever been presented to that effect. Many people have seen bees with eggs in their mandibles but no one has ever seen bees placing eggs in cells and it is quite likely that those bees with eggs in their mandibles were going to eat the eggs instead. As closely as honey bee behavior has been studied for so many years, it is highly unlikely that such a spectacular event as bees stealing eggs from another hive to rear a queen would have gone unnoticed. Once in a great while an unfertilized egg will go ahead and develop into a worker bee. In this case, had the cell containing that egg been changed into a queen cell by the workers, a queen would have developed, but this is an extremely rare occurrence and is hotly debated by some as to whether it is fact or fiction.

TWO QUEENS IN THE SAME SUPER

Q. Is it possible for two queens to exist in the same colony which is located in a single deep frame box?

A. Yes, there are occasions when two queens will exist in the same colony whether it's a small one or a large one. Usually, the two queens are related such as mother and daughter, rather than two unrelated queens laying side by side. The situation is due, normally, to the unit having raised a supersedure queen and as long as honeyflow conditions are good enough to warrant it, the bees will put up with both queens. This is certainly not the normal situation, of course. Generally, one queen is all a colony will tolerate. Usually when two queens do exist in the same unit, the bees will eliminate the older or the poorer one as soon as the honeyflow dwindles or whenever fall arrives.

CLIPPING AND MARKING OF QUEENS

Q. Will a queen remain clipped and marked for life?

A. Many times the paint will come off in less than a year — clipping is forever. Sometimes the paint does chip or crack off in part — sometimes the bees pull at the edges of the paint and eventually get it. Sometimes the queen uses her third legs to rub against the paint in an effort to remove it.

MULTIPLE MATINGS OR DRIFTING

Q. Authorities state that a young queen bee mates only when in flight, usually her first or nuptial flight, and when once mated, is mated for life. I question this. I have five colonies of bees. Two have Midnite hybrid queens, two had 3-banded Italian queens (one I just requeened with a Starline hybrid queen — no brood yet) and one has an apparently bronze Italian queen (caught in a late May swarm, origin unknown). All have been producing bees apparently true to their species until recently (late July) when both 3-banded Italian queens began producing bees that look like Midnite bees along with their regular brood. Apparently the bronze Italian queen of unknown origin, is producing what appears to be four species of bees in the one colony. Some appear to be Midnite and the remainder appear to be two different kinds of Italian bees.

In June, I noticed Midnite or Caucasian drones visiting all hives in my yard and I have my suspicions. Could the authorities be mistaken or is there another explanation for the change in offspring. I have checked the queen and they have not superseded.

A. We're sure the authorities are correct in stating that the virgin queen is mated in flight, and once she starts to lay is never mated again. She may be mated to several drones in the process and this may account for the different colored bees if they are being produced in the hives. It is also highly possible in this case that some drifting is taking place in your yard. This is not uncommon for one hive to pick up bees from several colonies.

QUEEN EXCLUDERS

Q. How is it possible for a queen to get through a queen excluder? I had one that got through the excluder three times. The last time I accidently killed her, probably for the good. The mystery still remains, because I had two other excluders exactly the same on two other hives and the queens never got through them. If you can answer this I would be glad to know why this occurred to prevent it happening again.

A. The difference in size between a queen and a worker is not great, so if by accident you happen to spread the wires of the queen excluder, even though it is hardly noticeable, it would be possible for the queen to pass through. It is also possible you had a small queen which could conceivably pass through the wires. The only other possible answer would be that you let her up there yourself in the management of the colony but this does not seem likely. You may have solved your own problem by killing the queen, even though it was accidental.

SCATTERING BROOD AND RAISED CAPPINGS

Q. When examining my bees in late fall I found scattered brood and many of the cells raised before capping. What is wrong?

A. It sounds as though you have a drone-laying queen or laying workers and in either situation, the colony needs some help. If your colony has a poor queen and she is laying drone eggs only, this would account for the raised bullet-shaped cappings. She should be replaced at the earliest convenience. If you cannot do that now, you could unite this colony with another weak colony by taking the lid off of the queen-right colony, placing a sheet of newspaper over those bees and then placing the drone-laying colony on top and recover the hive. The bees will gnaw through the sheet of newspaper and gradually become united.

MARKING OF QUEENS

Q. Some time ago you mentioned the International Color Marking of Queens. Is this determined by a pre-arranged chart?

A. The International Queen Marking color code has been developed by the Bee Research Association. The following code is recommended:

Year ending in 1 or 6 — white (gray)
2 or 7 — yellow
3 or 8 — red
4 or 9 — green
5 or 0 — blue

CROSSBREEDING OF BEES

Q. I would like to have a colony of Caucasian bees. My problem is, will these bees crossbreed if their hive is near a hive of some other breed of bees? How far should the Caucasian bees be from another colony of bees to prevent them from crossbreeding?

A. The hybrid Caucasian queen that you would order has already been mated. There will not be any crossbreeding going on unless the original queen is superseded and the colony raises a daughter from her. This daughter will then fly and mate with any drones that are available in the surrounding area. On these mating flights, the queen has been known to fly up to ten miles away during the mating procedure.

If you want to make certain that crossbreeding is not taking place, you should order a clipped and marked queen. Then, as you inspect the colony, you can tell when the original hybrid queen disappears and order a replacement.

LAYING WORKERS

Q. A friend of mine is sending for two queens for queenless colonies. Would it be good to take the combs a few yards away to shake off the bees to lose the laying workers?

A. If your queenless colony has laying workers in it, it should be a distinct advantage to take the bees some distance away and shake them out. The regular workers would find their way back to the hive, but the laying workers supposedly would not. There are many beekeepers who employ this method to successfully requeen a colony with laying workers. Some studies have also appeared that indicate that laying workers are quite capable of flying back to the hive so watch the requeened colony closely.

EGG-LAYING

Q. Do queens lay eggs the year round?

A. It would depend upon the climate. Here in the midwest, the queen usually tapers off on egg laying along in early November when there is no more need of new brood and will rest until the latter part of January or the first part of February when she will again start laying. The amount of eggs she then lays in the spring will be controlled by the amount of nectar gathering activities in the colony.

However, in California, Florida, and other areas where the climate remains warm, the queens may lay all winter on a greatly reduced scale, of course. Even in warm climates the queen may rest for a few weeks during a reduced honeyflow.

Fig. 91. A queen in the act of laying. (Photo courtesy, Ben M. Knutson)

QUEENS

Q. I have one colony of black bees that has not done well for the past two years. I was going to introduce a Starline queen, but when this queen came I found the colony had a nice new light yellow queen doing fine. Where could she have come from?

A. The yellow queen which appeared in your colony is probably from an egg laid by the original queen; she was probably mated to a yellow drone and as a result a percentage of her offspring will tend to be yellow in color.

RAISING QUEENS IN A QUEENRIGHT HIVE

Q. In raising queens in a queenright hive, about how far do we have to raise the brood frames and brood so that the bees in the raised part will raise queens and the queen in the hive will not travel up and destroy them?

A. In raising queens in a queenright hive, we would recommend that you place a queen excluder just above the lower brood chamber. In the brood chamber immediately above this queen excluder, you can raise up combs of eggs and larvae from below. This will draw the nurse bees to the upper chamber, while the queen excluder will confine the queen to the lower chamber. Your queen cells for rearing can then be placed immediately in this upper chamber, just above the queen excluder.

DEFECTIVE QUEEN

Q. I have an Italian queen that lays more than one egg in a cell in one of my hives. She was a new queen last fall and I raised her myself. This queen really showed her stuff. She laid from 2 to 5 eggs in each cell on 2 frames of brood comb. She had ample room to lay eggs on 4 to 5 other frames of brood comb. Knowing that this queen wasn't of any value, I pinched her head off and left her in the hive for the bees to find. I thought I would give them a chance to make themselves a new queen.

What causes a new queen to deposit her eggs in this manner when she has plenty of room? Will a new queen made from one of her eggs be all right or would it likely be like her?

A. Quite frequently a new queen will deposit more than one egg in a cell. This is particularly true in a small nucleus and frequently occurs as the queen just starts to lay. Any queen reared from such eggs should be perfectly all right.

However, the particular queen you describe in your letter was probably not a good queen. She was raised last fall and certainly could not be considered a young or new queen in mid-May of this year. The fact that she was laying anywhere from 2 to 5 eggs in a cell on two frames of brood, plus the fact that she had ample room to lay on 4 or 5 other frames, would indicate that she was defective in some way. There is one other possibility to consider. Sometimes a colony that has a failing queen will also develop laying workers. Laying workers invariably will lay large numbers of eggs in a cell. Such a colony also usually has several queen cells developing as the bees of the colony attempt to replace the queen. You did not mention that any queens were developing in your colony and we, therefore, assumed that it is simply a matter of a failing queen.

We recommend that you order a new queen and requeen that particular colony. Since it is probably a defective queen in the colony, we would not advise that you try to rear a new queen from her.

QUEEN CAGES

Q. I have two questions to ask you. The first is what is the formula for making the candy that is placed in queen cages?

The second question is how do you get the queen and worker bees into the queen cage after you have it ready for them?

A. Queen-Cage Candy (From Laidlaw & Eckert's "Queen Rearing"). "In the commercial production of queen bees, considerable difficulty has been encountered to secure a food that will sustain them without ill effects during the time they are confined to their cages. The best food per se is honey, and this was tried first. But it was discarded because the small combs were prone to leak from the package or smear up the bees. Then a combination of honey and powdered sugar was used, mixed together until it formed a firm candy that would hold its shape in hot weather. Honey is no longer used as a constituent of queen-cage candy because of the possible danger of spreading American foulbrood should any honey be used which came from an infected colony. (Some beekeepers still use this type of candy in their own cages, selecting a light-colored, well-ripened honey which they know was produced in an area entirely free of foulbrood).

"The candy which is now used universally for the commercial shipment of queens is the "Good" candy. It was first made in Europe by a Mr. Scholz, and so is also called "Scholz" candy. It is made by mixing invert sugar syrup and powdered sugar in approximately a 1:2½ or 1:3 ratio. The powdered sugar is first stirred into the sugar syrup. When the mixture becomes too thick to stir, it is kneaded as more powdered sugar is added, until a firm candy is formed that will not run out of the candy compartment in the queen cage.

"Beekeepers sometimes invert their own sugar solution by adding cream of tartar. But if this is not correctly done, the resulting candy will not be satisfactory. Much of the candy is now made from a commercial invert sugar known as Nulomoline, which is produced without the use of acid, and which is mixed with powdered sugar, as described. Some beekeepers also add a few drops of glycerine, which tends to prevent the candy from drying out. Sugar without starch is preferable to that containing a small amount of starch, but starch-free sugar is generally lumpy and the lumps hard to pulverize to a suitable fineness.

"Queen-cage candy should be left to stand for some hours before using in order to insure proper consistency. It should not be hard and dry nor soft and sticky. In hot weather, more powdered sugar is required than in cooler seasons. If properly made, the candy can be stored in an air-tight container for a considerable period.

'While queens and bees can be shipped through the mails for a period of a week or more when confined to queen-cage candy, they will not survive for a much longer period unless they have access to water to offset the dehydrating effects of the sugars. The small amount of starch in powdered sugar has little, if any, ill effects on the workers or queens. The main ingredient which is lacking in queen-cage candy, as made at present, is an adequate amount of moisture."

In regard to the second question, you should use the following method to put the queen and worker bees in the queen cage after you have it ready for them.

First, you should pick up the queen by the wings and head her into the opening of the queen cage. As she starts to go in, you should let go of her wings yet keeping your fingers behind her so she can't back out, and she will go on in of her own accord. After you have put the queen in you should place your finger over the opening to keep her in while you pick up individual worker bees by the wings and follow the same method of putting them in the queen cage. You should be careful, however, to make sure you grasp both wings of the worker bees. If only one wing is held, the bee can twist its body enough to sting you. After you have put five or six workers in with the queen, then you should close the hole of the queen cage by inserting a cork, usually, and your job is finished.

Fig. 92. A young hybrid queen being coaxed into a mailing cage for her trip through the mail to a beekeeper.

STORING QUEENS

Q. I am using a regular 10-frame hive, installing two tight dividers and using three frames in each section. From here on I need more instructions. I believe I have the wrong idea of this way to keep extra queens. Please explain the brood used or bees, also entrance or exits for each section, or any information you could give me.

A. There are many ways to keep extra queens in a beeyard. We prefer this method because a laying queen is always available and can be of some value in producing brood and even honey.

In this 10-frame hive, which is divided so that there is no movement of bees between sections, you will use one or two frames of brood and the adhering bees in each compartment. You will get better results if you move the bees a couple of miles to prevent drifting back to the original colony. Each section should be closed with its own inner cover. Introduce a new queen as soon as the bees have become settled.

You should put one entrance to the front and one on each side to prevent the confusion of entrances side by side. To make these side entrances, cut out a one-inch piece of the bottom board rail.

Check the colony at frequent intervals to see that bees are not crowded. Remove frames of brood or honey as space is needed, or feed when necessary. The honey and brood may be added to one of your colonies.

NUCLEI FOR HOLDING QUEENS

Q. Is it advisable to establish nuclei for holding queens?

A. Many beekeepers do keep a supply of queens on hand in nuclei and it is an excellent idea. The number of queens kept in each yard as spares is about 10 per cent of the total number of colonies in the yard. That way, requeening can proceed without delay whenever necessary.

We hold queens by taking a frame of brood and bees from each colony in a beeyard and placing them in a hive with a screened bottom. We take enough so that we have two frames for each nucleus. This hive of bees and brood is taken to the yard where the nuclei are to be made up. Two frames of brood are then placed in each section of a 10-frame hive body which has been divided into 3 parts of 3 frames each with entrances in different directions. (The divisions must be solid so that there is no movement from one section to another.) An empty comb or one with honey is used to fill the empty space. The entrance to each section is restricted and a new queen is introduced. The bees must be fed until they are well established. Once established they should be checked at frequent intervals and have surplus frames of honey or brood removed from time to time so the queen will have adequate space. The frames of brood which are removed may be needed either in other colonies for added strength or may be used to make increases.

Fig. 93. This hive has been divided into nucs. The entrances face opposite directions and the solid divider allows no communication within the hive.

These laying queens can usually be placed on a comb of brood after the old queen has been removed and the colony given a puff or two of smoke. The bees and queens can be observed for a minute or two to be certain the bees are not fighting the new queen. Then the comb is returned to the hive.

Fig. 94. One method of holding queens, a specially-rigged frame for holding queen cages, allows bees to feed the queens.

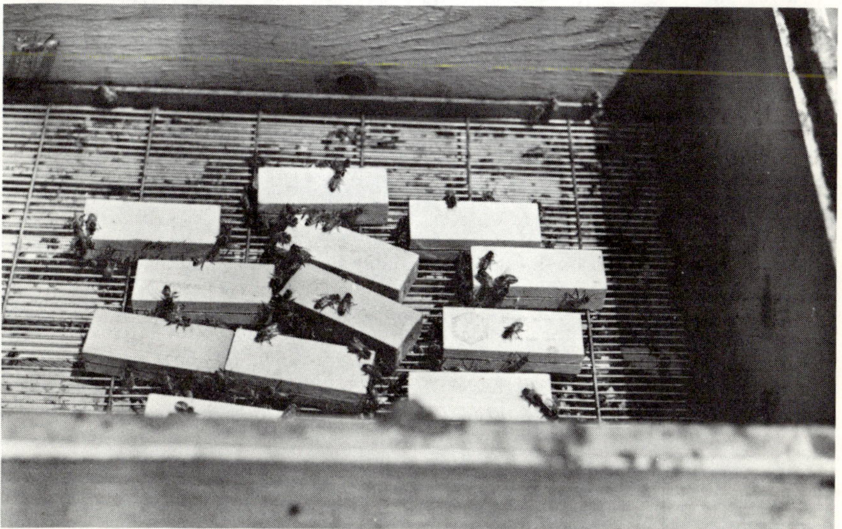

Fig. 95. A system of holding queens in a hive using a queen excluder and an empty super.

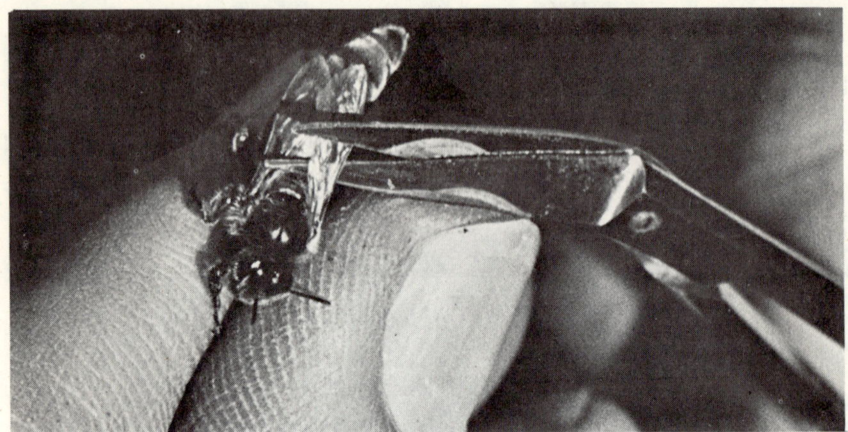

Fig. 96. Clipping the queen is a delicate operation. Most queen breeders will clip and mark queens for a small fee.

CLIPPING QUEENS

Q. Tell me how to go about clipping queens to help prevent swarming.

A. Clipping the queen's wings will not prevent swarming. It does keep the swarm low and if the queen is lost in the grass, the swarm returns to the hive. They will probably swarm again in 4 to 6 days when a new queen emerges but it does give you a second chance. Hold the queen between the finger and thumb at the thorax. Cut off a portion of both wings on one side. Be careful you don't cut a leg and ruin the queen's ability to move over the combs. Some beekeepers can clip the wings while the queen is walking on the combs but we don't try this. If you are killing the queens off each fall in requeening, we can see no reason to clip the wings. A package properly supered and tended should not swarm. One of the best measures in swarm prevention is proper supering to ensure added room as the bees need it, work ahead of the bees to keep them busy, and adequate ventilation.

TWO-QUEEN SYSTEM

Q. What is the reason for always leaving the old queen in the bottom part of a two-queen system?

A. The main reason for leaving the old queen in the bottom part of a two-queen system is based on the whole procedure. When the bees and combs of brood are placed in the upper body, the old bees will return to the old colony at the bottom. This leaves a preponderance of old bees in the bottom body. This would make it difficult to introduce a new queen to this group of old bees. On the other hand, the top unit will be composed of mostly young bees and it is quite easy to introduce a new queen to this age group of bees. A second reason for doing it this way is that as long as the old queen is in the bottom unit, there is no need to check her. It is easy, then, to check the acceptance of the new queen in the top unit.

REQUEENING

LOCATING THE QUEEN

Q. I want to take the old queen out of one of my hives but cannot find her. Would it hurt if I just put in a new queen? Is it all right to have two queens or will they swarm?

A. The problem you are facing is at the very heart of the whole requeening process. The inability to locate the old queen has thwarted beekeepers in requeening efforts for a long time. The art of spotting the queen while going over the combs in the colony can be developed with time and patience, but that doesn't do you much good now. As you suspected, the removal of the old queen is necessary for successful requeening to take place. If you just introduced a new queen, you would run the risk of having her destroyed by the workers or the old queen. Although occasionally two queens will lay side by side in a colony, it is normally a mother-daughter situation and not two unrelated queens.

One suggestion would be to take a piece of excluder and tack it across the entrance to the colony. Then, remove all of the combs from the hive, shake all the bees in front of the hive, replacing the combs inside. If the only way into the hive is through the queen excluder, the worker bees will be able to enter the hive with ease and after a short period of time, you should be able to locate the old queen on the outside of the excluder. We have used this technique ourselves here in the past, usually in association with the bee blower, but shaking and brushing the bees should work as well.

Fig. 97. Practice finding the queen on this comb. Learn to recognize the distinctive shape of the queen among the thousands of workers covering the combs.

HYBRID QUEENS

Q. When introducing hybrid queens, are any special techniques or precautions necessary or can introduction be made in a similar manner to that of regular queens?

I note, too, that two different hybrids — Starline and Midnite — are offered. From the literature I assume that the Starlines are of Italian stock and would therefore, be used in preference when this type of bee is in the apiary. What might result if the Midnite strain was used in a colony of Italian browns? I understand the Midnites are mostly Caucasian stock.

A. Your assumptions concerning Starlines and Midnites are correct. As you had assumed, the Starline is comprised of inbred lines derived from Italian bee origin. The Midnite, on the other hand, comes from mainly Caucasian background, although there is a bit of Carniolan stock in one of the lines. Introduction techniques used for the hybrid queens would be the same as for any other queen bee.

There would be no problem when introducing a Midnite queen to a colony of Italian browns. As soon as the Midnite queen is accepted and laying, it's just a matter of time before her progeny will have taken over the population of the hive. There will be no resentment on the part of the Italian browns for the Caucasian.

VICIOUS COLONIES

Q. I acquired four hives as a small hobby and did not always have time to tend them so my father kept the grass and weeds down. He could do this on 3 of the hives but the 4th would always run him out. If he got within 10 feet they would get violent although they never bothered me or my children.

There was an old gentleman close by who kept bees commercially for years. He never used a smoker but always had a corncob pipe and used long green tobacco. The bees never bothered him but he also never made a sudden movement.

A. It sounds as though you have a vicious colony that should be requeened with a gentle strain of hybrid bees such as Midnites. We don't know why the colony singled out your father for punishment except that perhaps your second paragraph about the old gentleman who never made sudden movements may give us a clue. Handling any strain of bees is best done with quiet, sure movements. Perhaps your father used sudden movements that were tolerable by the 3 hives but just too tempting for the colony that always chased him out. Although the theory is untested, we have had many beekeepers tell us that bees seem to be able to sense fear and insecurity on the part of a beekeeper and this may have been your father's downfall. Fear or insecurity may change the secretions of the skin and perhaps the bees were able to pick this up as a signal to attack. Whatever the cause, requeening with a gentle strain of bees such as Midnites should calm the colony down.

PAPER BAG METHOD OF QUEEN INTRODUCTION

Q. I have heard of introducing a queen using a paper bag. Can you give me any information on this?

A. The method borrows from the newspaper method of uniting bees in that a small paper bag is pricked with a darning needle or fine pointed knife so that there are tiny holes in the bag that the bees can later enlarge to release the queen. Place about 20 or so bees from the colony that is to be requeened into the bag and close and shake the bag for at least two minutes. Make sure you have moved away from the hive to shake the bag as the bees may not appreciate the sudden movements. When the bag is opened, the bees will be stunned and lying in the bottom of the bag. Separate the queen to be introduced from any attendant bees and drop her in the bag so that she is down at the bottom. Twist the bag shut and put a piece of string around it, part the center frames of the brood nest and hang the bag down between the frames. Suspend the bag from a thumbtack pushed into a top bar and the thread twisted around it. Close up the hive and leave for at least three days. The bees will begin to carry out scraps of paper within the hour but leave the bees and the queen to become introduced gradually.

CROSS COLONIES

Q. I have one colony (out of three) in which the bees have become very mean. They become very angry as soon as the brood chamber is disturbed. I would like to replace their Italian queen with a Caucasian. How can I get rid of their queen? How can I find her? I do have a queen and drone trap. Would inducing the bees to ball the queen do any good?

A. The easiest way to find the queen is to take out the frames one at a time and look for her. She is located usually in the brood nest near eggs. If you do the job carefully and with a minimum amount of smoke and confusion, she should be spotted easily. Some beekeepers will put an empty hive body under the full brood chamber with a queen excluder between. The bees are shaken off each comb onto the queen excluder and the frames are placed below in the empty hive body. The bees will all try to go through the queen excluder to the brood, leaving the queen and drones above.

It is also sometimes possible to introduce the queen into a queenless nuc or small queenless colony over a double screen placed over the colony to be requeened. After the new queen begins to lay, pull the screen. Usually the queen in the top body will survive. There is only one sure way to eliminate the old queen and that is to find her and kill her. We would not risk introducing the new queen into the hive without first disposing of the old queen.

QUEEN INTRODUCTION

Q. Recently while working in pollination experiments, I encountered a method of queen introduction with which I was not familiar.

A mixture of sugar water (1:1 sugar to water) is combined in a 1:1 ratio with vanilla extract. This mixture is sprayed on the bees in the hive or package and also on the new queen. The queen is then released from the cage and placed directly with the bees, regardless as to whether the queen traveled with the bees or not. Supposedly, the vanilla mixture destroys the scent of the queen and the sense of smell of the rest of the bees.

A. Actually, the use of vanilla and other strong scents for queen introduction is not new and works on the theory of masking the colony odor by which the bees can tell who belongs to the colony and who does not. By the time the bees have everyone cleaned up, they all smell the same and introduction has been accomplished.

RELEASING ATTENDANT BEES FROM MAILING CAGES

Q. What is the best way to remove attendant bees from the mailing cage before introducing the queen?

A. Release the attendant bees by holding your finger over the hole in the mailing cage and as a bee comes up, let her out. If the queen comes up, use your finger as a stopper. An easier and surer way would be to release the bees in a small cage or room. All of them will fly toward the light. Pick the queen off and put her back in the mailing cage and let the attendant bees go. Your beeveil could be used for such a cage.

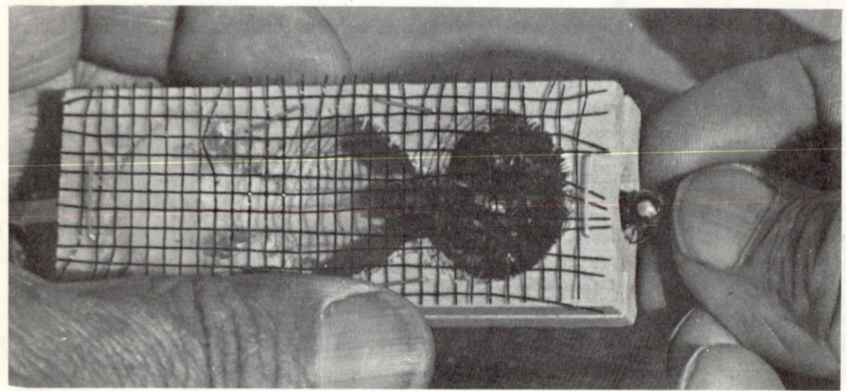

Fig. 98. Releasing the attendant bees prior to introducing the queen. Use your thumb as a gate to keep the queen confined.

REQUEENING

Q. This is my first year in beekeeping and I have a few questions. I hear people talk, each does things differently and gets results that please him, some I like and some I don't. Some I use and some I don't. What I want to ask is, do you believe it necessary to requeen each year to obtain maximum production and colony strength. Also, is it necessary to requeen each year to help prevent swarming?

I have read three books which are, *The Hive and the Honey Bee, First Lessons in Beekeeping,* and *How to Keep Bees and Sell Honey.* The first two listings say no, the third believes it necessary to requeen each year. I know your people were connected with writing the first two listings but even though I read them the thought hits me — why go into my hives and kill queens who are prolific layers and requeen them even though the young queens were of their mother's bloodline? It seems rather wasteful. Are old queens really that much more inclined to swarm, if given proper management, and if laying well — producing gentle, healthy bees?

A. There is nothing magic about requeening every year. In fact, many times such a practice can remove a very good prolific queen from a colony and replace her with a queen that has been reared under rather poor queen-rearing conditions and that harms your colony. We prefer to think that a colony should be requeened for reasons that beekeepers can measure. This would be because the colony is too cross or vicious to work with, because it does not winter well and do well in the spring, or because the queen is run down and generally unable to produce a large colony of bees. Sometimes these items are not easy to evaluate. In general, however, it is true that a colony with an older queen that is beginning to fail, will swarm more than a colony with a young queen. Also, in general, it is better to requeen a colony of cross or vicious bees than it is to allow such a colony to perpetuate itself by supersedure or swarming. The swarming line or the crossness may be perpetuated into the progeny of the next generation. Both swarming and vicious temper are inherited characteristics that do, indeed, pass down from one generation to the next and should be eliminated from your stock both for your own comfort, as well as for the ability to achieve larger crops in the future. Of course we are going to recommend the use of the Starline or Midnite hybrids. Either of these two types will give you greater honey production than average common stock as well as other good traits such as gentleness, quietness, and minimum swarming.

REQUEENING IN NOVEMBER

Q. Is November too late to requeen the older hives?

A. Generally, now is not a good time to requeen. It is far too late in the year. You could have done the requeening any time in September or the early days of October, but November is getting too late. The best thing you could do would be to wait for spring and requeen your colonies at that time.

REQUEENING

Q. I run my bees with a queen excluder over two ten-frame full depth hive bodies. This year I did not get my supers off until now, January and February, with below freezing temperatures.

I find some hives with large clusters of bees up in the supers, and I did not disturb them. I expect some of these hives will prove to be queenless, the bees having moved up and left the queen below the excluder. Would you advise me to attempt to requeen the ones that prove to be without queens, or would it be better to unite them with queenright colonies and divide them later when they have several frames of brood.

A. Your problem is a rather difficult one and brings up a point about queen excluders. Queen excluders are probably the most controversial piece of beekeeping equipment but they can be a valuable management aid if used properly. Many beekeepers put them on too early and become dissatisfied with them while in your case, they were left on too late and the results are equally bad. Had you been able to at least remove the queen excluders and leave the crop on, the bees would have wintered and begun spring build-up in fine fashion.

If conditions are such that you can examine the hives, remove the queen excluders and try to determine the true condition of the colonies, whether they have queens, if so, are they beginning to raise brood, and do they have sufficient quantities of food available to them. If your colonies have queen excluders on them and you are finding large clusters of bees up in the supers, the bees have probably moved up through the excluder in search of honey and have left the queen behind. If that is the situation, the queen is probably dead and the colony is queenless. The bees above the excluder will have sensed their queenless condition and you should be on the lookout for a laying worker if you find eggs above the excluder. It is highly unlikely that you have a new queen above the excluder as the only way the bees could have raised a new queen would be if the queen had been able to go through the excluder and lay in the supers.

If you have queenless clusters of bees above the excluder, we would suggest that you see that they have honey for survival and merely let them alone until the earliest date you can safely requeen in your area. Introducing queens and/or uniting colonies is almost too tricky in below freezing weather to try it.

REQUEENING

Q. If I find a queen that is not laying a uniform pattern what should I do?

A. Requeen the colony with a hybrid queen as early in the season as possible. A queen that is not laying a uniform pattern simply cannot lay enough eggs to produce the brood required to store any surplus during the honeyflow. The fact that she is not laying a uniform pattern of eggs would indicate that she is a failing queen and the situation is not going to get better.

A poor brood pattern can also be from matings which genetically are incompatible, wherein as much as 50% of the eggs may be inviable and never hatch. Such a problem could cause such a shotgun pattern of brood and there is no way to correct it short of requeening with a quality hybrid queen.

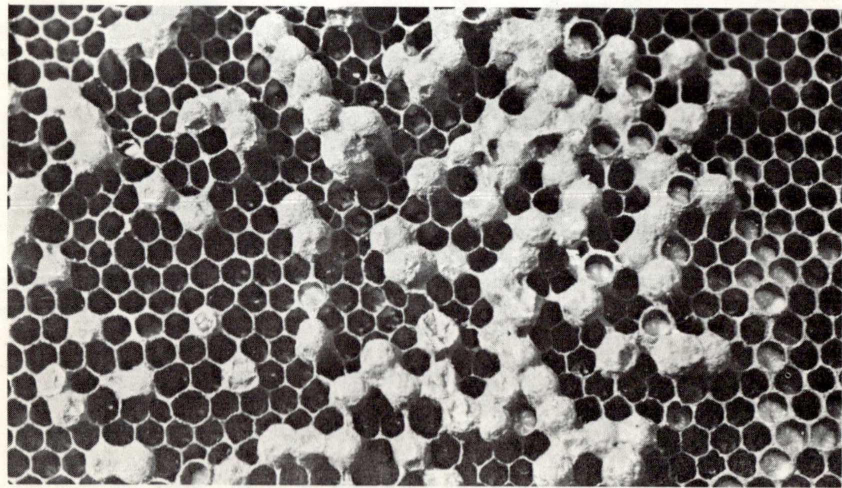

Fig. 99. Drone brood, the result of a laying worker.

REQUEENING THE "LAZY" COLONY

Q. I am a beginner and have three colonies of bees which I plan to keep as an avocation. I purchased these colonies early this spring as packages and each has a hybrid queen. Two of these colonies seem to be doing quite well during this honeyflow period. However, the third colony seems to be a "lazy" colony, and is considerably behind the other two colonies in their production. I am wondering whether it would be advisable at this point to replace the queen this fall and supplement the hive with another pound or two of bees early next spring?

A. Requeen the lazy colony as soon as possible, even though there are probably other factors which are influencing this colony's production. It just isn't strong enough. We doubt that package bees will be necessary if you get the colony headed by a young queen and strong enough to winter by fall. Requeen and feed if necessary to go into the winter with adequate stores.

REQUEENING

Q. I have two hives that have raised their own queens this summer and are no longer hybrid. When is the best time to try to replace these queens with hybrid queens?

A. In general, colonies of bees are requeened in early spring or in the fall, but given the right manipulations, colonies may be requeened any time of the year. The bee clusters are smaller in the spring and the fall, and this makes the task of finding the old queen somewhat easier. Regardless of the time of requeening, keep some general rules in mind. Remove the old queen prior to introducing the new queen. Some beekeepers operate on the theory that the new queen will automatically kill the old queen and this is not true. Likewise, do not leave the colony queenless — remove the old queen and introduce the new queen, using either the mailing cage or a push-in cage. Sprinkle a little sugar syrup over the top of the brood nest cluster immediately after requeening and just prior to closing the colony. This should not be overdone to the point that the syrup runs out of the colony, just sufficient to attract the bees within the colony.

The requeened colony should not be disturbed for at least a week to ten days. The new queen must have time to emerge from her cage, be accepted by the worker bees, and commence egg laying before the beekeeper checks to see if introduction was successful. The presence of eggs and young larvae ten days after introduction is sure proof of acceptance.

REQUEENING

Q. Recently I was discussing requeening with a local beekeeper with many years of experience. He told me when his new queens arrive he then kills the queens that are to be requeened, dropping them in the hive so the bees will know the old queen is dead. Then three or four hours later he releases the new queens directly on the frames or through the bottom entrance with very good results. He also said that if he waited until the next day, the bees would be working on queen cells and would not accept the new queen in this way. I have never heard of this before. What do you think?

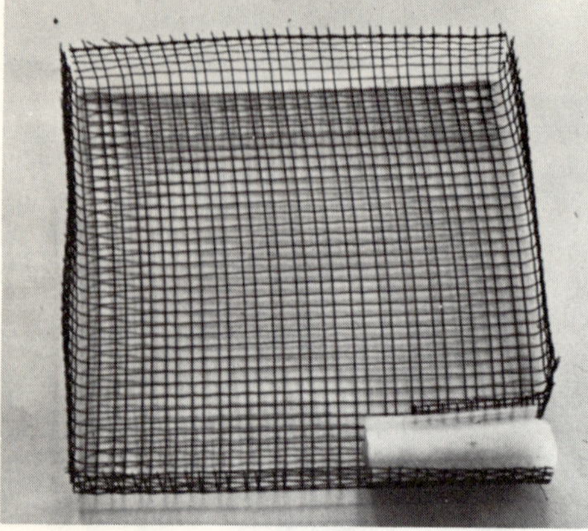

Fig. 100. This method of queen introduction uses the push-in cage. The workers will eat through the white candy which blocks the entrance to release the queen.

Fig. 101. The queen is placed on the surface of the comb and the cage is pressed into the wax to hold it in place. Be careful to allow room for the queen to move about inside the cage.

Fig. 102. By the time the workers eat through the length of the candy, the queen will have acquired the distinctive odor of the hive and will be acceptable to the bees as the new head of the colony.

A. If your friend has a system that works, he should stay with it, but it sounds risky to us. We would guess that leaving the dead queen in the hive would only hinder acceptance of the new queen. The bees don't need a body to prove to them that they have no queen.

There are many methods being used to introduce queens and almost every beekeeper does some little thing that makes his system work for him. We prefer to stick to the old standard ways of killing the old queen and immediately hanging the queen cage containing the new queen between the frames after the attendant bees have been released from the mailing cage. We have always had good results from this method. Your friend is correct when he says do not wait until the next day or you will have queen cells and rather poor chances of acceptance, but we suspect there is little to be gained from waiting three or four hours either. Any time a queen new to the colony is introduced directly, there is always a chance that she will be attacked and killed before she acquires the hive odor and acceptance from the bees as their queen.

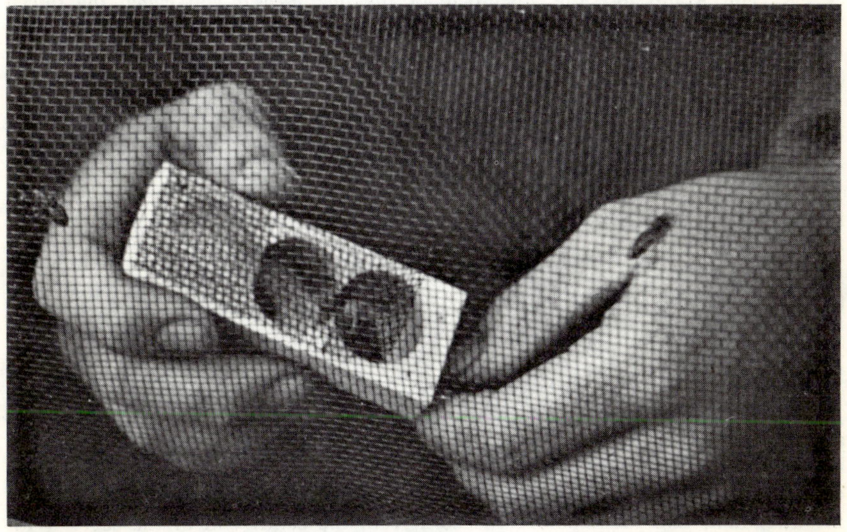

Fig. 103. Attendant bees will not be well-accepted by the hive bees and are being released inside a screened enclosure. Your bee veil will work well as a cage.

REQUEENING WITHOUT LOCATING THE OLD QUEEN

Q. How can I requeen a colony without finding the old queen?

A. We have tried requeening without finding the old queen by introducing the new queen into a queenless colony made of two or three frames of brood and some food. After the new queen has been accepted, her colony is united with the colony to be requeened by placing the new colony on top with a sheet of newspaper between. Usually the queen in the top colony is accepted. Chances of success with this method are as high as they are in introducing a queen to a queenless colony. This division or small colony could also be placed immediately above a double screen over the old colony to be requeened. When the top queen has begun to lay, the screen may be removed.

REQUEENING TO CHANGE RACES

Q. If you have a strong hive of Italian bees and you requeen it with a Midnite queen in September, how long do you think it would take for all the Italians to be gone? Also would any race requeen the same way? My aim is to put new blood in my hives.

A. We would expect it to take 6 to 8 weeks before all the Italians are gone from the hive, but you will notice a change in about one month. The same method of requeening may be used for any race or strain of bees.

REQUEENING FROM A NUC

Q. I am going to take one frame with the queen on it from a nuc this spring and put it in a colony that needs requeening. Will the bees kill the queen or shall I cage her? Shall I shake the bees off the comb or leave them on?

A. You would lose the advantage of the nuc idea if you caged the queen and it would not be necessary to shake off all the bees. The best way to proceed would be to transfer the whole nuc into the hive. First remove the old queen and the same number of combs from the hive that you are going to transfer from the nucs into the hive. The old queen can be kept in the nuc until a new queen is available or, if a new queen is available at the time, she can be killed and the nuc requeened. If you do not desire to reestablish the nuc at the time you requeen the hive, simply kill the old queen and move the whole nuc into the hive after removing enough empty combs to accommodate the combs of the nuc.

REQUEENING

Q. I have three colonies of bees that are queenless and without brood or eggs. Will they accept a queen now?

A. The queenless and broodless units should certainly be ready to accept a queen. However, since they have been queenless for some time, we recommend that you use either some sort of push-in cage or else the queen-mailing cage for introduction. If you intend to use the queen-mailing cage, first remove the attendants and then remove the cork from the candy end of the cage. With a nail or wooden match, force a small hole through the candy leaving a small hole less than an eighth of an inch in diameter. Then suspend the cage between two of the brood combs making sure that the bees have access to the screen side of the cage. They will quickly eat through the candy end of the cage and release the queen, but not before she has acquired the hive odor.

BEST TIMES FOR REQUEENING

Q. When is the best time of year for requeening?

A. Requeening can be practiced at any time of the year that it is needed, but it is much easier in certain times of the year than in others. During the late spring and summer, when the colony is at peak population levels, the number of bees in the colony may be so great that it is extremely difficult to locate the old queen. Therefore, fall, after the colony has cut back in brood rearing and the population level is down to that point at which it is ready to go into winter, or the early spring before the colony has had a chance to build up are the easiest times to requeen.

REQUEENING

Q. When requeening, how can one find the old queen? I have looked many times, but still haven't found one yet. Also, how long after killing the old queen should the new queen be installed? Last spring, after not being able to find the old queen, I installed the new queen by putting honey on the bees in the queen cage and inserting it in the entrance. The bees got them out of the cage all right but now I wonder what might have happened to them?

A. Finding the old queen is a matter of experience, looking very carefully, and a little luck on the side. Once you have seen the queen on the comb you will have little trouble as you will be looking also for the circle of attendant bees surrounding the queen and this will give you a larger target to look for. If you can get an experienced beekeeper to show you how he finds them, it will help, and it will also be useful to you to have your queens clipped and marked.

As a last resort, and every beekeeper has probably had to do this at one time or another, shake all the bees on top of a queen excluder. The bees can go down through the holes of the excluder but the queen and the drones will be found on top of the excluder. The recommended procedure is to place the hive body containing the frames and bees on top of a queen excluder which is placed on top of an empty hive body, slide the queen excluder to one side far enough so that as the frames have been shaken, they can be placed in the empty hive below. Keep sliding them over until all 9 or 10 frames are below. The bees should go down through the queen excluder to the brood below, leaving the queen above. The same method will work by screening the front of the entrance to allow only workers to pass through and then shake the frames of bees off in front of the hive, making all the bees pass through the excluder. When all the bees have returned into the hive, the queen and drones will be found on the screen. I personally do not like to disturb the bees this much but sometimes nothing else works.

The new queen should be introduced immediately after the old queen has been killed or removed from the hive. It would appear as though your new queen was not accepted. Putting honey on the queen cage has never proved helpful to me although I have heard from beekeepers who swear by the system. There are many methods of requeening but I try to follow these standard rules: 1. find and kill the old queen; 2. release the worker bees from the queen mailing cage before introducing the new queen; 3. hang the new queen in her mailing cage from the tops of the frames much as you did when installing the package of bees; 4. make sure that you have provided a starter hole through the candy end of the queen cage; 5. close the colony and do not disturb for three or four days to let the bees release the queen and accept her.

REQUEENING

Q. I have what is probably the strangest colony of bees in the world. They never make any honey beyond what they require to keep alive. I have supered from earliest fruit bloom to fall when the goldenrod comes in, but never any honey have I seen from this hive. Great masses of bees congregate on the apron of the hive and sit there with their tongues to the board, rocking back and forth from their front legs to the back ones, looking like an army of miniature vacuum cleaners. The hive has been requeened twice. The first queen disappeared in a few days; the second one was a Caucasian and seemed to be accepted but was superseded, and still the vacuum cleaners go on operating. A hive four feet away, from the same stock, made two shallow supers of honey last year. Any suggestions? I don't make a living from bees but simply to see if it can be done. I should like to see if this hive can be brought into production.

A. We have observed the cleaning motion of the bees which you describe and your colony should be requeened. We would suggest you requeen by uniting it with a queenright nuc or a small hive. Place the new colony on top of the bees you wish to requeen with a sheet of newspaper or double screen between. It might be insurance for acceptance if you kill the old queen.

Fig. 104. A virgin queen, newly emerged from her cell.

REQUEENING

Q. When a hive needs requeening, what vital signs appear and what should you look for?

A. Usually, requeening can appear obviously necessary through the egg laying pattern of the queen. If it is at a time of the year when the brood nest should be rapidly expanding such as in the spring, and the colony is lagging far behind in buildup, it can be suspected that the queen is old and not able to provide the strong egg laying needed by the colony, assuming that there are no shortages of honey or pollen. A second sign would be a spotted egg laying pattern with a queen missing large numbers of cells in her pattern. Of course, another sure sign is when the bees start producing supersedure cells themselves. Quite often the bees do know best and there is a good reason behind bees starting supersedure cells to rear a replacement queen.

ROBBING

ROBBING

Q. What is the standard way to clean wet supers that will not be replaced on the hives until the next year? What do commercial beekeepers do in this case?

A. It seems to be a toss-up between commercial beekeepers as to whether or not the supers are stored wet or whether they put them on to have them robbed out dry. Some feel that the wet supers bring the bees up into the supers quicker in the spring when they place them back on. However, this is a bit messy for storage. Trying to super at the beginning of the flow with wet supers from the year before is a good way to get wholesale robbing started in a yard. Those that do try to clean the supers up usually take them out and set them off in a beeyard in the fall allowing the bees to rob out the wet honey remaining in the combs. It can also be done by placing the super on top of the hive, preferably above the inner cover so there is just a small opening for the bees to get up and do the cleaning. This would result in a less severe robbing situation.

Fig. 105. A colony that has been killed by American foulbrood is being robbed out. Such a hive should be removed from the beeyard to prevent the spread of disease through robbing.

HOW TO STOP ROBBING

Q. How does one stop robbing?

A. The best method to stop robbing is to start preparations before the robbing is started. When the nectar flow is over and nothing more is coming into the hive, there is always danger of robbing. Many beekeepers unconsciously start the process by opening the hive during this danger period, going to the bottom of it, taking out frames and placing them near the hive, or extracting frames and placing them out in the open during that time. If you have to go into a hive when there is no nectar coming in, do it in a fast fashion, get the hive closed and watch the entrance for a short while. If you see bees landing on the entrance board and other bees grabbing their wings, you know that robbing has started. Put an entrance cleat in the front of the hive, then watch for a while. If you see what appears to be strange bees going into the hive and two bees coming out together, fighting, you know robbing is in process. Pull some grass and throw it over the opening in the entrance cleat and observe for a while. If this does not stop the robbing, close the entrance completely and leave until dark, then remove the block, leaving the entrance cleat in. Look at the hive the following morning and you will possibly see some snoopers around, but most likely the robbing will be stopped. If you must check into the complete hive, try to do it while a nectar flow is on in the spring or fall.

Fig. 106. Robbing in process. The entrance guard bees will require some help to protect the hive.

Fig. 107. The use of robber cloths to discourage robbing during hive examinations.

DISCOURAGING ROBBING WITH COVER CLOTHS

Q. I also seem to have problems with robbing in my beeyards and I would appreciate some help.

A. Robbing is a condition with which all of us may have to contend at one time or another. This is simply a different expression of the bee's honey gathering instinct. When the supply of nectar shuts off, they continue looking and looking for a new source. The guard bees in very weak colonies, queenless colonies, or those with drone-laying queens may be overcome easily, and whatever honey they have will be packed off by the robbers.

During robbing season, the beekeeper can foil the robbers to a greater or lesser degree, depending on how careful he is in his manipulations. Following a flow, there are usually burr combs built between the combs in one super and those above or below. There may be others built between combs in a super. Removing a super or frame breaks these burr combs allowing the honey to drip. Scrapings of such burr combs from the tops of the frames should be placed where the bees cannot get to them. Honey which drips onto the ground will be found by the foraging bees. When they return to their own hive with this new-found honey, the others rush out to get their share. Some of the bees are bound to find the broken bits of burr comb on the tops of frames in the open hive. These robbers, in turn, arouse the guard bees in the hive so that if allowed to continue, they are ready to pounce on anything — including the beekeeper!

Several precautions may be taken to minimize the initiation of robbing. Setting supers into lids placed upside down will keep much of the drop off the ground. We use cover cloths — somewhat like dish towels — dipped in water. One cloth can be placed over the super, or supers, setting on a cover. Being wet, they do not blow off easily. In some instances we have put a small amount of benzaldehyde or other chemical repellent in the water. Either acts as a bee repellent.

For working in the brood area of the hive, it is best to use two cloths, one on each side. Each can be moved so that the comb you wish to remove is exposed. Once the comb is out, one or the other of the cloths can be pulled over the area again. If it is necessary to remove several of the combs, it might be best to have an extra empty body, or super, sitting close by to place the combs in. This, too, should be covered with a cloth.

The old saying, "An ounce of prevention is worth a pound of cure," is most certainly true concerning robbing. So, it is much better to use extreme care in handling supers which may drip honey and to use cover cloths before robbing starts.

Should robbing get started, and it is necessary to continue working in the yard, it is probably best not to smoke the entrances of the colonies too much since this excessive smoking can demoralize the guards, leaving these colonies vulnerable to the robbers.

The commercial beekeeper with many colonies in one yard and a financial interest in getting his work done with a minimum of trips to each yard, will undoubtedly have much more of a problem with robbing than will those with a few colonies. Such a beekeeper realizes that the ultimate defense against robbing is to maintain strong queenright colonies and to work carefully so that the robbing is not man-induced.

ROBBING

Q. I am a newcomer and I ordered bees last year and had very good luck. I also picked up a swarm late last summer but they seemed to be rather slow and had only two frames of honey at the close of the season. I was feeding them hoping to keep them through the winter but a warm spell came and my large hive of bees robbed the weak hive, leaving an empty hive. I don't know if my weak ones joined up or went off and died as there were no more dead bees in the hive. Now I have ordered another package to arrive in April. What shall I do to keep my older bees from robbing my new package when they arrive? I have only a small backyard.

A. In spite of the problems you had with that late swarm, it sounds like your first season was quite successful. The problems that did develop with the later swarm are not surprising, as they undoubtedly didn't get enough of a chance to build up a good strong colony and collect enough stores for winter.

Probably the easiest way to solve your problem with the new package would be to keep a substantially reduced entrance in the package colony until they have build up enough strength to defend themselves. This can be accomplished with an insert that may be purchased or you could use a board or stick that would fit across most of the entrance. I would leave an entrance area of only about two inches wide and perhaps ½ inch high until the package is soundly established. Then, if you notice the other bees trying to rob them out, you can try spreading grass loosely over the entrance. Covering the entrance this way slows down the entry of the robber bees and gives the guard bees of the package colony a better chance to defend themselves and their colony.

STINGS

AMOUNT OF VENOM IN AVERAGE BEE

Q. I was wondering if you could tell me how much venom is in the average working bee?

A. We did some digging on the amount of bee venom in the average working bee. The literature points out that a small amount of venom is already present in bees two days old. However, the amount builds up until about 18 days when production is no longer continued. There was quite a bit of variation in the literature on the amount of venom in a single bee. Back in 1951, Dr. Hoenig of the University of Minnesota had a review in the report of the Iowa State Apiarist. In this one, he mentioned that the venom of the adult bee would be around 0.3 mg. Then in the 50's Benton and others of Cornell developed a device for collecting venom from bees. Using large quantities of bees, they were able to collect only about 0.1 mg of venom per bee. Any way you look at it, it's not a very large quantity. Incidentally, Benton found that the venom they collected from the adult bees ran 35 to 50 per cent solids.

STINGS

Q. First let me state that I am a novice in beekeeping. I have just started and have a lot to learn. I am a school teacher. The other day (in October) I took a group of students out to watch while I robbed a colony. Several of the students were stung. One boy was stung near the left eye. The next day his eyes was swollen shut. During the noon hour I gave each student a small spoonful of comb honey. Within thirty minutes his eye was opening and two hours later it was open. In your opinion, did the honey have anything to do with this or was it likely that his system just cared for it? I am very interested in finding out about this.

A. I was unable to find any reference to the use of a spoonful of comb honey as a treatment following a reaction to a sting. I would like to suggest that such demonstrations to a group of children should be done only when the conditions are optimum for having gentle bees in the hive. Any given hive will respond differently to varying conditions, weather as well as the honeyflow conditions, sometimes from day to day.

Before they are taken to see the bees, the children should all be told how very important it is to remove the stinger immediately by scraping the stinger out rather than picking it out with two fingers. Squeezing the sting out with two fingers is like squeezing the bulb on the end of an eye dropper — it forces the poison from the poison sacs into the flesh. However, if the stinger is scraped off, even though a small portion of the stinger should remain in the flesh, the poison sacs will be removed with only a minimum of the poison ever getting into the flesh.

Such things as hot compresses, or baking soda, dampened and placed on the point of where the sting entered may help, probably from a psychological viewpoint. However, there is one product called Sting-Kill which will neutralize the effects of the acid if the material is placed on the flesh as quickly as possible after having been stung. Sting-Kill is available from Dadant & Sons, Inc. and it might be wise to have a supply on hand for future field-trips.

STINGS

Q. My 16 month old son wandered to a beehive with a stick and ended up with 20-25 stings. I took his clothes off and applied vinegar with a scraping motion to try and remove the stingers. Then I applied soda water. To my surprise he did not swell at all. He had a high fever the next morning and was very tired. I have been told it would be dangerous for him to get a number of stings again. Is that true? In this case, if the bees had continued to sting him would he have gone into shock?

A. We don't know why your son did not swell as a result of the stings, except that he is apparently not allergic to bee venom. His reaction seems normal and your treatment was not wasted in that it probably removed many of the stings with a minimum of poison being injected into his system. There seems to be some doubt as to the value of external medication but you should check with your physician to get a treatment for bee stings. Usually an antihistamine is effective.

EFFECTS OF STINGS

Q. I have six colonies of bees now; two headed by Dadant hybrid queens, two by 3-banded Italian and two by Golden Italian. A few weeks ago, while I was working with them, two or three stung me. In a few minutes I became hot all over and began to itch and burn all over and again yesterday as I was working with one of the Goldens, two bees stung me, one on the right arm above the elbow, the other on the right little finger. I paid no attention to the stings but went ahead working with them. But again in a few minutes I became hot and began to itch and burn all over from the bottoms of my feet to the top of my head.

Up until these two occasions, I had been stung many times with no ill effects at all. Now I don't want to get rid of my bees if I don't have to. So what can I do? Is there a medicine I can take before going to work with them that will prevent this burning and itching and make it safe for me to work with the bees?

Of course I only keep bees as a hobby and although it's only a hobby, I would still like to be able to handle bees that would put up at least a little surplus. I have had a few of the yellow Italian bees that were real gentle but the ones I have now are about like the three-banded. I have never tried the Caucasians. What do you recommend? Should I get rid of my bees? Could I requeen with a more gentle bee say in September?

A. Our suggestion is that you see your doctor and follow his advice. It won't be safe for you to keep bees if you are becoming more sensitive to their stings. As a rule, people who start to become allergic to bee stings get progressively worse as time goes on. Many doctors will prescribe some drug, usually an antihistamine. There is also a series of shots which will desensitize some people to the effects of bee stings. These may be of help to you and your doctor would be able to administer the shots.

If there is a most gentle bee, it is the Midnite, a hybrid bee bred from Caucasian lines. Of course, even this bee has stings so your best course is to see your doctor and investigate your reactions to bee stings further.

SUPERING

COLONY MANAGEMENT

Q. Is it necessary to have more than two supers per colony?

A. Yes, many times in a very fast nectar flow a good colony of bees will have nectar scattered through as many as four supers and not have any of it cured and capped, and ready to take off. Therefore, if you do not have the extra supers, you will lose part of your crop.

SUPERING

Q. I have 6 hives of bees and I gave them two boxes for winter. Can I add 2 more boxes at one time in May? A party told me that is what he does and that is all he adds. Where should I put the inner cover? Leave it on the 2 boxes or put on top of the 4 boxes when I add the two? Also would like to put on a few small boxes for comb honey, 1-lb. boxes, as I have some supers for comb honey filled with 1-lb. boxes. Where should I put them?

A. In May, you can add two or more boxes or supers at one time and it may be plenty of space, but we usually find that bees should be checked and supers added as needed. They seldom get away with only two supers. The inner cover goes on top of the stack under the cover.

Bees, as a rule, do not efficiently produce both comb and extracted honey at the same time. They produce one or the other in a single colony. Bees should be confined to one hive body during the honeyflow if they are to produce the best comb honey.

SUPERING

Q. I would like to know why bees may not fill out the second super but will fill the first?

A. There could be any number of reasons why the bees would fail to fill out the second super, but most likely what has happened is that the honeyflow has ceased or essentially ceased before they had a chance to get at it. This, of course, would be the most common answer. Other possibilities would be that your field force may have been destroyed by a pesticide application and, therefore, you had no bees to bring in the remaining nectar.

PARTLY FILLED FRAMES IN UPPER STORIES

Q. At the end of each honey season each of our hives contains a great many partly filled honey frames. Most of these frames will have from 5 to 50 cells filled with capped honey. We will have so many partly filled frames in the higher supers that in the fall it becomes difficult to cut the colonies down to two or three stories. How can we force the bees to remove this honey to the lower stories?

A. You could place the frames containing the cells of honey together in some supers, put one of these supers on top of a hive with an inner cover between the hive containing the bees and the super containing the honey, the hole

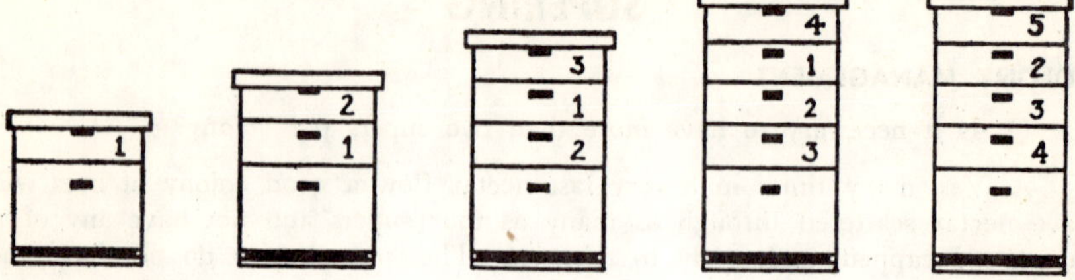

Fig. 108. Diagram showing the method of supering. The supers are numbered in the order in which they are given to the bees. When the colony is ready for a third super, it is placed on top of the other two supers. Before giving the third super, the first super should be nearly full and the second one at least half full. The supers should be removed from the hive as they are filled.

in the inner cover being left open. The cappings on the honey should be broken with a cappings scratcher and the bees will carry the honey down. All supers should be removed from the colonies first, and the inner cover and super with the honey placed directly on top of the hive.

If you are sure all of your bees are free from disease, the cappings on the honey can be broken and the frames set outside where the bees can get to them. They will rob the honey out but that is a good way to spread disease so be certain your bees are free of disease first.

It would seem to us that you put supers on your bees too rapidly. According to your letter, you have several supers on each hive with scattered cells of honey in them. The second super should not be put on the bees until the first super is nearly full of honey.

VENTILATION

Q. Every afternoon when it is hot, the bees form clusters outside the hive. Do they need more ventilation? Later when it gets cooler they go back into the hive.

A. The bees likely hang on the outside because the colony is heavily populated and overcrowded. They need shade, ventilation and some attention or they will solve the situation themselves by swarming. First, completely remove the entrance block to give the bees full access to the hive entrance and also to aid in ventilation. Then let's look at your supering to see if the bees need another super added. This would ease the overcrowded situation and give those bees something to do in drawing the foundation in the frames of the new super. If more ventilation is required, the inner cover could be moved back from the front edge about a quarter of an inch, giving more air space. The supers could also be staggered slightly front and back about bee space width to further increase the movement of air through the hive. If additional ventilation is needed you might raise the hive itself off the bottom board by inserting small blocks of wood at the corners. All of these techniques will aid the bees in keeping the temperature of the interior of the hive near normal but be careful about leaving the hive in such condition for long periods of time as you will have opened up more spaces than the bees can defend and you may encourage robbing. You might also consider a light shade such as placing the hives in the fringe area of the shade of a tree with light open branches.

SUPERSEDURE

SUPERSEDURE QUEENS

Q. I understand that it is probably necessary with the failing of a queen, that the workers would take charge and replace the queen if the beekeeper had failed to notice the problem. Now, if my bees are trying to supersede their old queen, would it be wise for me to take a healthier queen and replace the failing queen with it?

A. You have sensed the correct thing to do in the case that the bees are appearing to supersede their queen. This would be an excellent beekeeping practice to remove the failing queen and replace her with a strong prolific young queen. Otherwise, you will probably lose anywhere from 7 to 14 days of egg laying in the changeover of the queens while the new queen gets mated and prepared for egg laying.

QUEEN SUPERSEDURE IN PACKAGE BEES

Q. My experience with bees tells me that queen supersedure as described by several good books on bees and the queen supersedure of package bees is two different things with the same name.

In all cases of package bee supersedure I have had (which comes about the time the first brood is ready to emerge) the queen seems to have mysteriously disappeared before any attempt to construct queen cells was made.

Most authorities on supersedure will say that the old queen is tolerated until the new one is about to or has already taken over. What is the difference in the two? Were the queens of the package balled, died of disease or what?

I am planning to get a few packages, from the South, this spring. Have you any advice or suggestions on what to do to discourage this trouble?

A. Supersedure merely means the bees are attempting to requeen the colony by themselves, but there seems to be a difference in the two cases. In a normal colony, which has a failing queen, the bees will build queen cells and will tolerate both queens for a while; but in the case of package bees, the queen often seems to disappear after only a few eggs are laid. There has been some work done on this problem which indicates that, in many cases, this is caused by Nosema disease, a protozoan, which affects the digestive system of the bee. Although there is no positive cure for the disease, some good results were obtained when Fumagillin or Fumidil-B was fed to overwintered colonies from which package bees were shaken. Through experience, you will find which package bee shipper gives you bees that will live. You may try those produced by shippers who feed Fumidil-B.

QUEEN SUPERSEDURE

Q. Early this spring one hive started making queen cells on each side of the center frames. As I would find them I would cut them out. In a couple of weeks the queen had disappeared and as I have removed the queen cells, there was no supersedure queen raised and I had to order another queen. I had queens that were marked and clipped. Should I have left the queen cells to produce another queen or should I have destroyed the cells as I did and ordered another queen? I don't know if the queen swarmed or just died, as I don't see any decrease in the number of bees in the hive. This morning I checked my other hive and it had the same story, 5 or 6 queen cells on each side of the 4 or 5 center frames and no queen bee. I destroyed them all as I checked each. From results, all queen cells destroyed and the colony has no queen. Some of the cells were capped. Incidentally, this hive produced 180 pounds of pure white honey from catsclaw and mesquite. The hive that lost the queen early this spring only produced 50 pounds. I am now wondering if I shouldn't requeen each fall, as my two queens that were one year old, failed during the spring honeyflow.

A. It sounds as though your bees were trying to supersede or replace their old queens. You probably should have removed the old queen and allowed one of the queens to emerge from their queen cells but, you would have had no control over the new queen's mating and she would not have been the true hybrid quality queen that you had before. Another clue is that if the colony only produced 50 pounds of honey, you were in need of a new queen anyway. You might also have removed both the queen and the queen cell and requeened the colony with a hybrid queen. The hive that produced 180 pounds of honey sounds as though it had a rather good queen and a supersedure queen from that hive might well have been a good one.

Requeening in the fall with tested hybrid queens is going to be the safest route to follow since there is virtually no aging effect on your queen for having gone through the winter with the colony, and your colonies will have the advantage of young, vigorous hybrid queens in the spring to get them off to a good start.

SUPERSEDURE

Q. In a small colony the bees superseded their queen. About three days after the queens emerged, there were eggs in some drone cells but no eggs in any of the worker-sized cells. Since drone eggs are never fertilized, were these eggs laid by worker bees or the virgin queens?

A. If you are certain the old queen is not alive, the eggs were probably laid by laying workers. The virgin queens usually don't mate until they are about seven days old and seldom lay sooner than three days after mating.

SUPERSEDURE

Q. Our hive has been doing fine except that the bees constantly build queen cells while we continue to destroy them. We presently have two deep supers and one shallow super. The two deep supers are full of comb, partly filled with honey and partly with larvae and eggs. The shallow super has foundations in it and was placed along with an excluder about six weeks ago. Still the bees have not built on the foundation.

We destroy the queen cells to prevent the bees from swarming before next spring when we plan to divide the colony. Could overcrowding be the cause of the queen cells because the two deep supers are packed with bees (the shallow super has bees in it but it is far from full)?

We want to divide them in the spring to again have two colonies but should we wait till spring or divide them now (midsummer)? If we should divide them now what would the procedure be?

A. Remove the queen excluder from between the super and the top body. Sometimes bees are hesitant to build new combs on foundation especially if the flow is dwindling. Some colonies are more easily discouraged trying to go through the excluders than others and in this case, the excluder acts as a barrier to them.

Although it is very difficult to try to analyze what might be wrong without having seen the colonies themselves, the queen in your hive is probably failing and is unable to deposit the number of eggs which the colony thinks she should. Perhaps at this time of the year it would be a good idea to just let them go ahead and raise a new queen and supersede the old one. You could do this by tearing out all of the queen cells excepting one or two good ones.

We would hesitate to divide the colony now since the fall honeyflow would be the only thing left for them to build up on. If you let the colony supersede the queen and they get a fair fall flow to build up on, they should be an excellent colony to go through the winter. By spring this colony should still be in good shape so that it could be divided into the two colonies you would like to have next spring. Be sure to check for winter stores and if they are low, provide additional food. You will also have to order an extra queen for next spring.

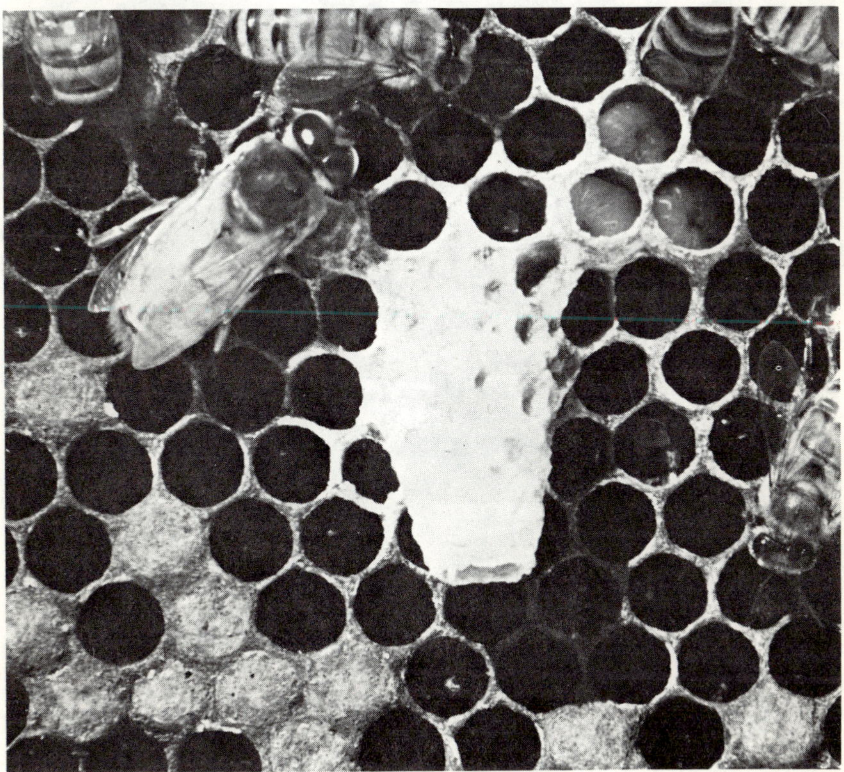

Fig. 109. An unfinished queen cell constructed on the face of the comb. (Photo courtesy, E. R. Jaycox)

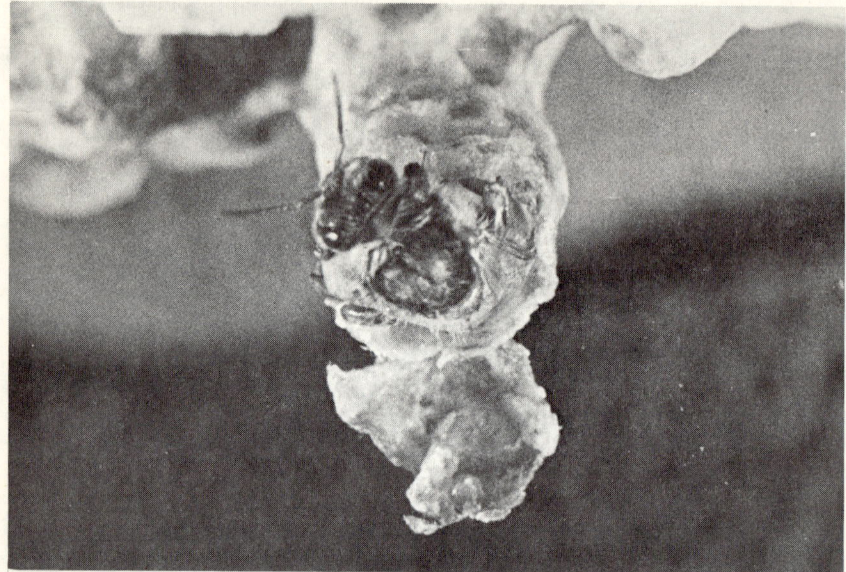

Fig. 110. A virgin queen struggling to emerge from her cell. (Photo, Encyclopaedia Britannica Films, Inc.)

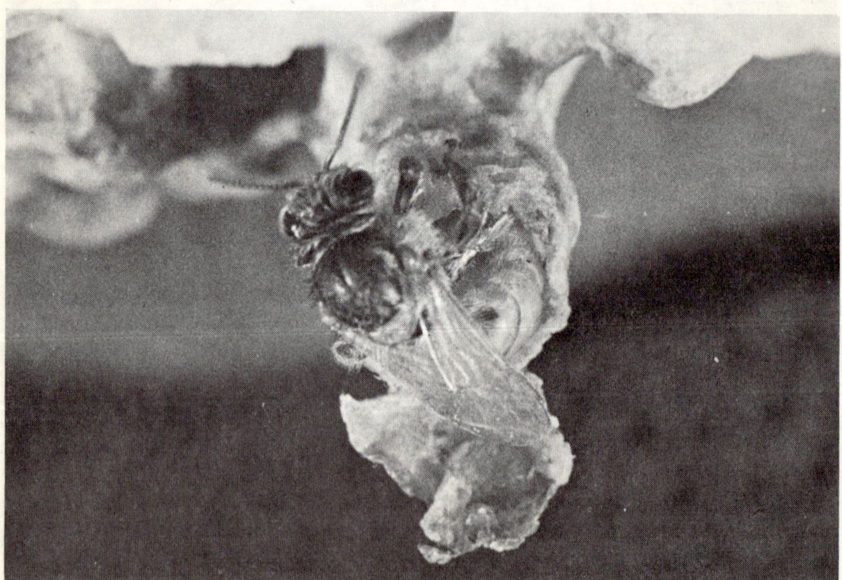

Fig. 111. Almost completely out of the cell. If she is the first virgin to emerge, she will have a good chance to become the head of the colony. (Photo, Encyclopaedia Britannica Films, Inc.)

SWARMING

SWARMING

Q. Where is the queen when the bees swarm into a tree and how long will they stay before leaving?

A. When a swarm clusters away from the hive, for example, on a limb of a tree, the queen goes with the swarm — unless clipped or injured. A number of times when we have attempted to find the queen in a swarm, it seems that she would be near the top and towards the center of the cluster of bees.

Once the swarm has clustered on a tree or such, it may remain there from 15 minutes to an hour or even a whole day. Sometimes when the weather is bad, the swarm may hang there several days. But the normal condition is for the swarm to break up and move on to their new home within a few hours of having settled.

There are times when a swarm may, for some reason or other, decide to stay in its position and they will start to build combs. The queen will start laying in the combs, and soon you have an outdoor swarm. This probably happens when the old queen which left with the swarm was still rather heavy from egg laying and finds it difficult to fly. Another factor which may affect this is the honeyflow. If it is quite heavy, then the bees find it easier to produce wax and build the new combs necessary for this type of colony.

Figs. 112-113. For some reason, known only to the bees, they have decided to build their combs in the open. Although it is a good opportunity to study the natural structure of honeycombs, the colonies will probably not survive the winter as both combs were built in cold-weather areas.

RANGE IN SIZE OF SWARM

Q. How many bees would be in an average swarm?

A. The size of a swarm would depend on the situation. A very small handful of bees could swarm in a last after-swarm or in a small nuc that swarmed. The swarm pictured here was one of the larger ones we've seen in some time and it weighed about twelve pounds. Assuming approximately 3500 bees to the pound, this swarm would have contained about 42,000 bees. You understand that these are estimates as most swarming occurs without the benefit of scales and the beekeeper's main thought is to capture the swarm rather than weigh it.

Fig. 114. A large swarm that has settled temporarily on a low hanging branch. This swarm will be easy to hive.

YOUNG QUEENS FOR SWARM PREVENTION

Q. I am told that I can prevent swarming by keeping my colonies stocked with young queens. Is this reliable?

A. There are many causes of swarming: overcrowding, improper ventilation, and supersedure impulses to name a few. Supersedure of the queen is Nature's way of replacing a failing queen and by keeping your colonies stocked with young queens, you will have forestalled one of the swarming impulses. However, you will still have possibilities of swarming from the other causes unless your management program also solves these problems before they reach the point of swarming.

Since requeening with young queens alone will not solve the entire swarming problem, keep a close watch on all of the potential problems a hive might have during a working season and requeen only when it becomes necessary. Many queens will give good service for two years.

DRONE EGGS IN SWARM CELL CUPS

Q. Does a queen ever lay drone eggs in those small swarm cell cups at the bottom edge of the frames or must one always assume that when they have eggs in them that queens are being raised?

A. Under normal circumstances, we doubt very much that the queen would ever lay drone eggs into the small cell cups which are frequently found on the bottoms or ends of the combs. We would assume that when a normal queen lays eggs in these cups, that the colony is preparing to swarm or, in the case of a failing queen, preparing to raise a supersedure queen. Under abnormal conditions, such as an old queen which has no more sperm left, an unmated virgin queen which has started to lay, or in the case of laying workers, it might be possible for a queen to lay drone eggs in the cell cups.

CLIPPED QUEENS AND SWARMS

Q. What happens to a clipped queen when she tries to swarm and must return to a hive full of virgins?

A. Under normal conditions, when a colony tries to swarm with a clipped queen, the bees would stay with the queen even though she was able to get only a short distance away from the hive. We have noticed many times that the swarm will leave the colony and fly to a nearby limb to light in a normal manner. However, they soon realize that the queen has not followed them to this limb and become quite nervous, eventually returning to the hive. If the queen is on the ground nearby, there is usually a small group of bees with her. We have returned the queen to the hive after cutting the queen cells and killing any virgins which might be out. If, however, the bees tried to swarm a couple of times like this and failed, the next step for them would be to swarm with one or more virgins.

VIRGIN QUEEN FLYING OUT WITH A SWARM

Q. Will an unmated queen fly out with a swarm?

A. A virgin queen will fly out with a swarm, especially if the queen of the parent colony is clipped and cannot fly. Under such circumstances, a virgin queen may join the swarm. At other instances, virgin queens may fly out to mate and they may be joined by small swarms. These usually return to the parent colony after or before the virgin queen has mated.

G. H. Cale, Sr. states in *The Hive and the Honey Bee* that in very populous colonies, about a week after the prime swarm leaves, more bees may swarm out with virgin queens, apparently when the virgins take their mating flights. There may be several virgins in each of these swarms and this type of swarming may continue until the population of the colony is reduced to a low point.

SIGNS OF SWARMING

Q. Is there any way I can tell by the bees' actions when they want to start swarming?

A. An experienced beekeeper could sometimes tell by the action of the bees whether or not the colony is likely to swarm. Usually there is more drone activity, the bees seem to fly a little less and act more lazy. Hanging out in front of the hive may be a sign of impending swarming, but it also indicates other conditions such as overcrowding, lack of ventilation, or hot, humid weather. To be on the safe side, check the colony every eight days or so for queen cells.

PREVENTION OF SWARMING

Q. I am a retiree and plan on keeping a few hives of bees. Would you please tell me how to prevent them from swarming and also what to do when they are about to swarm? Would you also tell me what kind of bees to start with?

A. I'm afraid that no one has the answer as to how to prevent all swarming. The following conditions seem to contribute to swarming, so if you correct these conditions, you should have the minimum amount: (1) poor queens result in supersedure; (2) crowded conditions for brood and storage of honey; (3) poor combs with too much drone comb; (4) colonies which do not go into the supers fast enough.

What can you do? Be sure the colonies are strong enough to go into the supers. Requeen every year or two, or whenever you find a failing queen. Add supers before the bees need them, rather than after. Don't use a queen excluder until after the bees are well established in the supers. You will have fewer problems with extracted honey production than with comb honey. Use only good combs so that the amount of drone brood, which is of little value to your colonies, is at a minimum.

Most beekeepers use Italian bees but this is a personal thing. The Caucasians have some fine points and are worth a try. Why not try both Dadant Starline and Midnite hybrids and see which you like better.

INSPECTING COLONIES FOR SWARM CELLS

Q. How often should a colony be inspected for swarm cells? Would it be wise to destroy drone cells when it is possible as a swarm control?

A. A colony should be checked for queen cells every 8 days during the period before the honeyflow or the swarming season. It is a waste of time to destroy drone cells. You could eliminate the problem of drones by using a good queen and good worker combs.

QUEEN EXCLUDERS TO PREVENT SWARMING

Q. I was wondering why beekeepers don't simply put on a queen excluder on the bottom of the hive to prevent swarming by preventing the queen from leaving the hive.

A. The method which you described in your question is used sometimes by beekeepers to prevent swarming. However, we do not advise using a queen excluder for this purpose for several reasons.

First of all, the queen excluder can become clogged with dead drones and inhibit the entrance of the worker bees. Also, even if the excluder is kept clean, it still does provide an obstruction which the bees must pass through and thereby could lower your honey yield. Sometimes an abnormally small virgin queen will be able to slip through the queen excluder and swarm anyway.

We think it would be much better for beekeepers to take preventive measures by providing the colony with enough room during the honeyflow to prevent them from swarming. Even if a queen excluder is used, this will not prevent the colony from building queen cells during crowded conditions.

THE ALLEN METHOD OF SWARM PREVENTION

Q. What is the Allen method for swarm prevention?

A. We went back to the January, 1960 issue of ABJ to let Dr. C. C. Miller answer your question:

"When the honeyflow is well started, I go to each strong colony, regardless of whether the bees desire to swarm or not, and remove it from its stand, putting in its place a hive filled with empty combs, less one of the center ones. Next, a comb containing a patch of unsealed brood about as large as the hand, is selected from the colony and placed in the vacant place in the new hive; a queen excluder is put on this lower story, and above this a super of empty combs, this one having an escape hole for drones; and on top of all, an empty super. A cloth is then nicely placed in front of the new hive, on which the bees and queen are shaken from the parent hive, and the third story is filled with the combs of sealed brood and brood too old to produce queens, and allowed to remain there and hatch, returning to the working force.

"This is really the Demaree plan, which was given to the public many years ago by G. W. Demaree, a prominent Kentucky beekeeper at that time. Mr. Allen has varied it by putting a frame with some brood in the lower story, whereas Mr. Demaree had only empty combs or combs with starters in the lower story."

DEMAREE METHOD OF SWARM CONTROL

Q. What is the Demaree method of swarm control?

A. The brood of the colony is examined and all queen cells are destroyed. The hive is then removed from its bottom board and a body containing one comb of unsealed brood, eggs and the queen is put in its place with the remaining space filled with empty combs. A queen excluder and the remainder of the brood and the bees are placed at the very top. The colony still has all of its brood, and the queen is in the lower body with a free brood nest.

In 10 days, examine the brood combs in the top hive body and remove all queen cells that may have been built in the interval. In 21 days, all of the brood will have emerged in the upper body and it will be used for honey storage, while bees will be beginning to emerge from new brood in the lower body, so that a continuous succession of young bees is maintained. Except in unusual seasons, it is seldom necessary to use the Demaree plan more than once.

THE TWO-QUEEN SYSTEM AND SWARMING

Q. I have been using the two-queen system for several years. Last year nearly all the colonies swarmed. All but one or two of the old queens got lost when swarming and the swarms went out again with virgin queens. How can I improve on this situation?

A. It may be that the two units were united too early. They should not be united until the flow has developed to the point where each single unit is well started in nectar collection. If united before this, there are too many bees with too little to do, so swarm cells are started.

Assuming that your swarming occurred in springtime, it may have been due to inclement and intermittent weather which confined the bees to the hive and in such circumstances they need something to do such as drawing comb foundation. Some beekeepers make a practice of giving supers of comb foundation to draw the worker bees out of the brood nest and to give them something to do. However, it sometimes seems that the bees do not read the same instruction books as we do and there are times when weather and other conditions result in their swarming regardless of what we do. There are also some years when swarming is worse than in others.

SWARMING

Q. Last week I found the queen (clipped and marked) on the ground with about 50 bees around her (not a swarm). I picked her up, put her on the entrance and she went in. I then put on a deep hive body for more room and extra food supply for next winter. I understand many in this area use the two deep hives for brood rearing and then later add shallow supers with a queen excluder. Why did the queen leave the hive? Also, when should I put the original hive brood chamber which is about full on top of the one I just added?

A. It does sound as if the bees attempted to swarm with the queen, but inasmuch as she was clipped, she could not fly with them. You did the proper thing by placing her back into the hive, but you might have done well to check to make sure there weren't any ripe queen cells in the unit. If there were and the old queen was not allowed to get at the new virgin, the hive could still swarm with the young virgin. Incidentally, one deep hive body is probably not sufficient to handle the brood rearing. We would recommend that two deep hive bodies be used for the brood chamber and then your honey supers placed above that. This gives more versatility for you in the spring as far as manipulating the hive to stimulate buildup, and, as you mentioned, gives more food storage space for the winter.

The normal sequence of events in the hive is for the bees and queen to move upward in the hive with the egg laying as the queen needs the space. Usually, when the honeyflow begins to come on strong, the worker bees will force the queen back down into the brood chamber again. However, in earlier spring she may go up through the entire hive, leaving the bottom completely empty. In advance of this happening, many commercial beekeepers practice what is called hive reversal. Both bodies are removed, the top one placed on the bottom board and the other on top. Then about two weeks later, they'll go back and re-reverse as necessary. This keeps plenty of egg laying room above the queen's head without letting her get clear up to the top of the colony.

PREPARING FOR A SWARM

Q. Last spring I acquired a swarm of bees from a friend. Due to the lack of equipment and a lack of know-how on my part, they swarmed again before the end of the season. I happened to be on hand at the time and it was a beautiful sight.

In anticipation of a swarm this season, I would like to be ready with a prepared hive. How many frames of foundation should I have in the hive?

Also, about ¼ mile away there is a bee tree (probably my swarm), too inaccessible to capture. From a disease point of view, should they be destroyed and what should be used?

A. It would be an excellent idea to get a hive prepared ahead of time. We assume that you are using the standard 10-frame hive. If the swarm that you catch this spring should be a large one however, it would be fine to use a full set of frames and foundation in the hive. It would be even better if you had some frames of drawn comb.

It is doubtful that there is any need to bother the swarm in the bee tree. Should it become diseased, the colony will probably die. Should this happen, then it might be wise to close the hole which the bees are using as an entrance.

DECOY HIVES FOR SWARMS

Q. In late June I prepared an empty hive quite some distance from the two hives I have, in anticipation of a swarm. The swarming took place at a time when I was away. And to my surprise I found them a few weeks later in the hive I had prepared. The shallow super that I had placed on the hive was full of comb and was about two thirds full of honey. Was this just luck or did I do something that encouraged them to do this?

A. It is almost like E.S.P. in that they knew what you were thinking or planning on. This is a purely natural response from a swarm of bees to move into a hive like this. In fact, many beekeepers frequently leave "decoy" hives in or around a bee yard so that during the season stray swarms may build in them. We have also heard of using a small ball of cotton placed in the back of the hive with a little anise oil sprinkled on it to attract the bees. This is something that you may wish to do again in other years.

NOISE TO CAPTURE SWARMS

Q. I have read of bees being calmed with sound one-half octave higher than the middle C note on a piano. I have also heard old beekeepers tell of bringing down a swarm by pounding on a dish pan. Maybe if they had the right size pan that gave the right vibrations per second it would work. What would you say to that?

A. We understand that some work has been done on bees' reaction to sound, but don't believe it had any practical value.

We have also heard of swarms being brought down when the beekeeper beats on a dish pan, but have never put much faith in the idea. The beating on a dish pan, as far as we know, came about as a result of an old English law which stated that the beekeeper must let everybody know that he is following a swarm to keep it in his possession. We would guess that beating a dish pan would serve to attract the required attention, but we have no verification that it had any effect on the swarm.

REMOVING SWARMS FROM BUILDINGS

Q. I am writing regarding two swarms of bees in a building. They are about six feet from the ground. What would you suggest to remove them? I was thinking of covering all the outlets but one and fix a wire screen with a bee escape to the new hive. Now the question. What do I do from here? What about a queen? These bees have been in the building for years.

A. Your idea will work, but you must either put a queen, queen cells, or a comb of eggs in the hive next to the bee escape so the bees will be able to rear a queen. It takes about 30 days to get all the bees out of the building. It is possible to then have the bees rob the honey out of the house.

CATCHING SWARMS

Q. I was always wanting bees. At last I bought two colonies which will swarm and I don't know how you catch the swarms. I wonder if you will help me by telling me how it is done? I'm bothered about it because I don't know how or when they will swarm. In fact, I don't know anything about bees but I am very interested.

A. Books have been written on swarming and its control; so you see it would be difficult to explain in one letter. We suggest that you do purchase some good books on beekeeping such as *The Hive and the Honey Bee* and *First Lessons in Beekeeping* and that you do some reading.

Bees usually swarm before the main honeyflow. In this area, we expect most of our problems in May and early June. Queen cells will appear in the brood chamber to indicate their intention to swarm. The bees (about half of them), the old queen, and some drones will leave just before the new queen emerges. When they cluster on a limb or brush, it is fairly simple to hive them. Place the hive under the swarm and shake the bees so the bees fall in front of the hive. They should walk into the hive without coaxing. Some old combs in the hive will make it more attractive.

Some swarming can be prevented by keeping the colony headed by a good young queen, giving the bees plenty of space for brood rearing and for storing honey. Partial shade and good ventilation may also be of some value.

HIVING A SWARM

Q. In August I found a small swarm in one of my yards. I killed the queen and they went back in. Since then I haven't been able to find a queen or any sign of one. I put in a frame with eggs and brood but they didn't start cells. I tried to introduce a queen but they killed her. In October when I covered them for winter, they still had quite a few drones. Could they have a virgin and are keeping the drones to mate with her next spring? Do you think killing the queen and letting them go back to the parent colony would be all right?

A. The colony in question either has a queen or will never have one. Usually a colony which has a queen will neither accept another queen nor will it produce queen cells even though you put in a frame of eggs. Once in awhile we find colonies which will not accept a queen by any of the usual methods. In this case the best procedure is to place the colony on top of a queenright colony with a double screen or sheet of newspaper between the two colonies. The bees will chew through the newspaper to become united. The double screen should be removed, if used, in a week or so to allow the colonies to unite.

The presence of drones could mean that the colony has a virgin, but it could also mean the colony is queenless. It's perfectly all right to kill the queen in the swarm and let the swarm go back to the parent colony, but you should try to find out why the colony swarmed and correct the condition. Some beekeepers make it a practice to requeen all swarms to try to improve the stock.

CROSS COMBS FROM A SWARM

Q. I have a colony of bees which was established from a swarm. The bees were dumped into a hive without any foundation and left to shift for themselves. They built cross combs. Last fall after the main honeyflow, I put on a deep super of wired foundation. This spring I found the bees had drawn the foundation into combs. I then put the top super on the bottom and the bottom on the top in hopes that the bees would move down but they wouldn't. How could I move the bees down so I could remove the top super?

A. If you allow the bees to become crowded, the queen will probably go into the bottom hive body with the good combs. You can then place a queen excluder between the two bodies and take off the crossed comb super in about a month, after all the young bees have emerged. We usually do it the other way; we place the good combs full of honey on top for the winter. When the bees are found in the hive body in the early spring, the crossed combs are removed. It is possible to smoke the bees out of the crossed comb into the one with the good combs.

FAILURE TO WORK IN THE SUPER

Q. Last year we hived a swarm of bees and gave them one super. When we examined the colony in late fall the bees had not stored any honey in the super although the lower body was full of honey. Can you tell me why they did not work in the super?

A. Perhaps there was not sufficient nectar available for your bees to store any more honey than they did. Also, bees do not like to go up into a super of foundation. It is often necessary to move some frames of brood up into the super from the brood chamber, providing the super is of standard depth. The frames of brood in the super entice the bees to move up and go to work. You might have an old queen in this colony if it has not been requeened since the swarm was hived. If so, she is not raising enough brood to make a populous colony with the good field force for bringing in nectar. Your bees should be inspected often and in swarming season, queen cells should be cut out. Try requeening your hive and see if that helps.

HIVING A SWARM

Q. I recently captured a swarm of bees and placed them in a new hive with proper frames and foundations. My questions are: (1) After three days I inspected the hive to find most of the bees clinging in a swarm on the inside cover. Does this indicate that I have possibly lost the queen? (2) Is it true that the queen with a swarm is an old queen and should be replaced? (3) Is it possible to save this swarm if it is queenless by introducing it into another hive of bees that does have a queen or should I requeen?

A. In regard to your first question, we suggest that you bounce the bees into the body and close the hole in the inner cover. You should also check the colony and make sure that the queen is still alive and in good condition. If you cannot find her or any evidence that she has been laying, then you should immediately obtain a new queen for the colony. This time we would suggest that you have her clipped and marked if you have not done so in the past. This facilitates the finding of her on the combs.

Yes, you are right in your belief that the laying queen is usually the one that goes with the first swarm. Subsequent swarms, if any, will have virgin queens. However, it is not necessarily so that the queen is past her prime for egg laying. Therefore, you should wait and see how she lays before you think about replacing her.

In regard to your last question, we would suggest that if the swarm is rather small that you go ahead and unite it with one of your other colonies. However, if the swarm is rather large and in good condition, we would suggest that you requeen the swarm to save it.

MAKING A SWARM STAY IN A HIVE

Q. How do you make bees stay in a hive when you hive them?

A. We assume that you are referring to catching a swarm. Sometimes it is difficult to keep the bees in the hive if all of the frames have foundation in them rather than combs. If, when the swarm is being put into the hive, the queen can be found and one wing clipped, this will prevent her from flying away again. Sometimes the queen will leave the new hive and when she does, the colony of bees goes with her. One thing which you can do to help keep the bees in the hive is to make certain that there is a feeder full of sugar syrup at the time the bees are hived. This will keep them from being hungry. Many times a colony of bees will leave if they are in a starving condition.

Fig. 115. Hiving a swarm. The branch has been cut and is lowered to the ground.

Fig. 116. The bees are docile and are easily handled without fear of stings.

Fig. 117. The swarm is placed on the ground before a hive.

Fig. 118. They will soon occupy the hive, giving the beekeeper another chance. (L. E. Strader)

EXCESSIVE SWARMING

Q. I reduced my colonies to 41 standard hives and last fall left one or two full supers on each. Also I have been requeening in July for several years with three-banded Italians. Then in March I had a large swarm and each day thereafter one to six swarms ending May 29th. In all, I hived 79 large swarms. Also 33 swarms left because I had no hives.

Why this unusual urge to swarm? How can I prevent it? Was it because of too much honey left in fall? Requeening too often?

A. You must have set some sort of record in collecting 112 large swarms from 41 colonies. Do you think that some of them might have been bees from other sources?

I believe much of this swarming could have been prevented by adding supers earlier. This is one of the problems which arises when so much honey is left with the colony over winter. One and a half or two-story colonies work well here, but we find some advantages in a little early feeding of sugar syrup to stimulate brood rearing.

If your colonies build up too quickly, you might divide the colonies to help reduce the swarming. Conditions might also have been unusual and the same problem may not come up again in years.

We doubt that either too much honey or too frequent requeening will cause the problem of excessive swarming. It would seem as though you had too many bees with too little to do and that lack of space in the hive was the main problem.

AFTER-SWARMS

Q. I am a beginner with only six colonies, two with Starline queens and four with dark Italian queens. At the beginning of last spring, I had two colonies, one with a Starline, the other dark Italian. The Starlines swarmed once, the swarm weighing about 9 pounds. The dark Italians swarmed three times, once in late April, again in early June and again July 15th. The swarms weighed 8, 10, and 15 pounds and yet the hive is still full of bees and emerging brood. Should I attempt to stop these after-swarms?

A. You should try to stop not only the after-swarms, but also the prime swarm. It is best to keep bees from swarming because the queen, a large portion of the bees, and part of the honey goes with the swarm. Also, there is a period when the colony is left without a laying queen. Swarming may be controlled in part by early and adequate supers, good queens and combs.

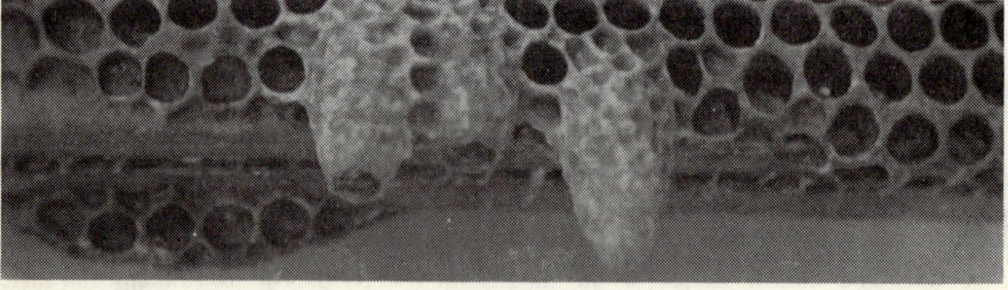

Fig. 119. Tell-tale swarm cells constructed along the bottom of the comb.

TRANSFERRING BEES

REMOVING BEES

Q. Can I transfer bees from a box hive to a hive with full foundation?

A. If you take the bees and queen from a box hive and shake them onto full foundation in a new hive, they may desert the new hive that day or the next. It is best to put a frame of drawn comb in the center of the foundation to attract the queen. If you have no drawn comb to put in the hive, take a piece of straight comb from the box hive, cut it about the size of a brood frame, place it in the frame and tie cotton twine around it to hold in place. In a few days the bees will attach this comb to the frame, gnaw the twine off the comb and carry it outside. Later this frame could be moved to the side and eventually done away with. Anytime you have comb that is irregular, move it to the side of the hive and let bees store honey in it as they will not have brood in the outside frame under normal conditions.

REMOVING BEES FROM LOG HIVES

Q. How does one get bees moved out of a log or box hive?

A. Generally, by attaching a standard hive to the entrance in the log or box hive and making sure that the bees must come and go through the standard hive to get to their combs in the log or box hive, the bees will become weaned, so to speak, to the standard hive.

In one case a beekeeper had a log sawed off square, some three to four feet long with the bees inside it. We told him to nail a board over the bottom part of the log so it would stand upright, then take a piece of masonite larger than the size of a hive body, cut a hole in the center about 6 inches square and place the masonite over the hole in the top of the log. Place inch strips around the masonite to fit three sides of the hive body and nail them. Then put the hive body on the strips over the hole with drawn combs or foundation, put the cover on and the result will be a hive sitting on top of an upright log. The only exit the bees now have is through the hive body and by being forced to use this hive body as a passageway to their combs inside the log, they will eventually move up out of the log into the hive body, the queen will start laying and finally all that will be left inside the log will be some old combs. You can then remove the log and work the bees into whatever type of management is desired. When coaxing the bees into the upper hive body, give them plenty of room so they will not crowd the queen back into the log. Also plug any holes that they may use as an entrance other than through the hive body. This method can also be used on old box hives that do not have removable frames.

TRANSFERRING BEES

Q. I am a beginner, and last June captured a swarm of bees and put them into an apple box hive where they remained and have been very busy. The box was bottom up and set upon a detachable board, with a slot for the bees to pass through. I wanted to transfer the bees to a regular hive, so in August asked a friend who knew something about bees. He obligingly filled a regular 10-frame hive with six frames of foundation, removed the apple box from its bottom board and set the box on top of the regular hive with nothing between them. He said the bees would thus have to pass up through the regular hive to get to their combs in the apple box and would probably move down into the regular hive. This worked out apparently well, as the bees have all remained and enter via the regular hive.

I saw my friend again last week and asked him about what to do next. He said about October to just remove the apple box, dump it as is on the ground beside the regular hive, and put a cover on the latter. The bees would promptly gather up all their belongings from the apple box and carry them into the hive.

Now I know that the apple box is festooned with honey combs, but whether the bees have moved any of it down into the hive is a secret between themselves and I don't feel to confident about dumping all that winter store onto the ground. So I would appreciate knowing the correct way to transfer my bees from their apple box to the hive, so that I can eliminate the apple box.

A. We usually do the job of transferring a little differently from the way suggested, although your method may work in California. We put the good hive body with removable frames on top of the box hive. When the queen is found to be laying in the new hive, the old box is placed on top over a queen excluder until all the brood has emerged (21 days). The old hive is then removed and the honey and comb rendered. The best place to put any new foundation, if you expect it to be drawn into comb, is above the brood nest.

I would not do this transferring in the fall, but rather in the spring when there is a honeyflow. Your bees may starve this winter if you take away their food at this time; they certainly would in the midwest.

It is highly unlikely that only the bees from the box hive will get the honey. It seems to me that all the bees in the area might come in and help themselves.

TRANSFERRING BEES

Q. I would like to know the best method of transferring bees from one hive to another, and which is the best time of the year to make the transfer. Would it be all right to transfer them to a used hive?

A. While bees can be transferred from one hive to another at almost any time of the year, it would be best to avoid cold weather and times when the bees are robbing. Spring is an excellent time because the colony is comparatively small. This would make it easier to locate the queen, thus minimizing the risk of crushing her while handling combs.

We suggest that the colony of bees be set just off its present stand and the empty hive be set in its place.

Then, starting at one side of the colony, remove the combs, one by one, and place them in the new hive in the same order. Any bees which fly off the combs during this disturbance will return to the new hive, thus creating a minimum of disturbance to their flight pattern.

There would be no objection to using a hive which had been occupied by bees before if you are certain there has been no disease in that colony. If there is any doubt about this, we would advise scorching out the hive, using a bottle gas torch or something similar. Any wax or propolis areas should be melted down and scraped out with a hive tool.

UNITING COLONIES

COLONY MANAGEMENT Fig. 120. How to Unite Colonies by the Newspaper Method.

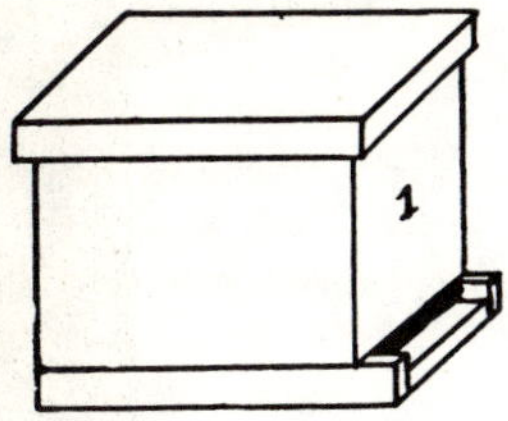

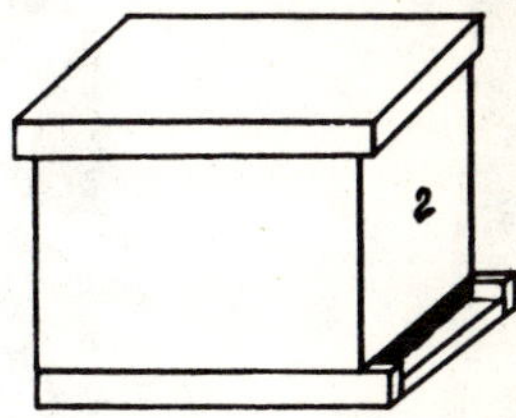

A. Assuming colony number 1 is stronger than colony number 2, bring colony number 2 over beside number 1.

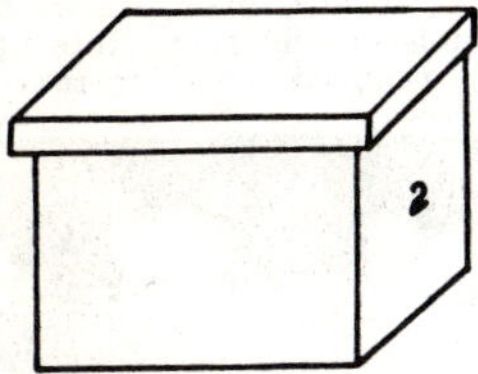

B. Remove the bottom board from colony number 2.

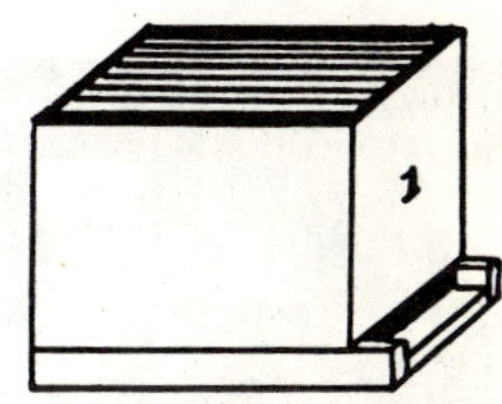

C. Remove cover from colony number 1 so frames are exposed.

D. Unfold a single thickness piece of newspaper and place it on top of the frames of hive number 1. Cut several small places with a sharp knife on the newspaper as indicated by the slash marks on the diagram. This will give the bees a start at chewing their way through to be united with the other colony.

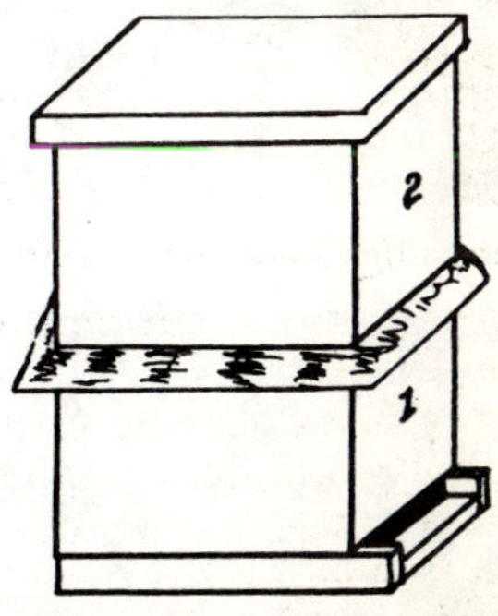

E. Set hive number 2 on newspaper so that the two bodies are directly over each other. Some of the newspaper will stick outside. It can be cut off or left to be removed later. The two united colonies, side view, should now look like this.

F. The bees gnaw away the paper and gradually mingle together as time goes on, and so unite peacefully without excitement. After the newspaper has been removed by the bees, the remainder may be removed by the beekeeper after several days, and the extra combs resulting from the union can be disposed of as desired. It is not necessary to find either queen unless there is a preference for one of them. Then the undesirable queen may be found and disposed of before the colonies are united.

Fig. 121. Preparing a colony for winter:

A. Feeding 2:1 sugar syrup to dispense drugs and provide adequate stores.

B. Insulite board over inner cover to serve as moisture-releaser. Note upper entrance and reduced lower entrance.

C, D. Installing commercial cardboard winter pack.

E. Colony ready for winter minus upper entrance.

F. Hole bored in cardboard pack to correspond to upper entrance in B.

WINTERING

PREPARING COLONIES FOR WINTER

Q. Our six hives are beginning to feel the cold weather, (New Jersey), and we are wondering which is the best way to keep them warm through the winter months until late spring.

A. There are undoubtedly many beekeepers in your area who do not bother to do any special hive wrapping for the winter. This would probably be unnecessary in your case also, if the bees are well protected from the wind. Another major factor is to assure the colonies of proper air drainage. An ideal situation would be to have some sort of a windbreak for the colonies on the north and west sides, and facing the colonies toward the south or east. A slope leading down from the entrances of the hive would be ideal.

Probably the best assurance of proper wintering would be adequate winter stores in the fall. If the colony does not have enough honey left on it, the feeding of sugar syrup can be accomplished quite readily. For late in the fall, or for early spring, probably the best sugar feeding formula would be two parts of sugar to one part of water. It will be necessary to heat the water to get the sugar into solution. The sugar solution can then be fed to the colonies in a friction top tin pail with tiny holes punched in the lid. This would then be inverted over the cluster. An empty super shell could be placed around the feeder can, and the cover put back in place. This places the warm syrup directly over the cluster and they can readily take it up.

PREPARATION OF WINTER STORES

Q. During late August one of my colonies had a shallow super filled with partially capped comb. I decided to wait until the combs were more completely capped before removing the super from the hive body. Two weeks later the combs in the top super were empty. I can only guess that the bees carried this surplus honey down to the main hive body. The possibility of swarming cannot be ruled out, but is not likely. Is this common late in the season?

A. The condition which you have described is quite common in the latter part of the season. Earlier, with plenty of nectar coming in, the bees had spread the honey over a wider area. This allows them to evaporate the water out of the nectar more easily in preparing the honey for storage in the cells.

In the meantime, the honeyflow probably slacked off or quit completely. Then, as you have guessed, when the honey was ready to be stored in cells, the bees moved it to the empty cells in the hive body below. This is actually positioning the honey in preparation for winter usage. It is unlikely that the colony swarmed since swarming is unusual this late in the season.

WINTER MANAGEMENT

Q. My question concerns proper winter (Minnesota) management. On September 17th, my two-story colonies, with ventilated cover and bottom board, weighed 166 pounds. Since then they were fed 2 quarts of two-to-one sugar syrup containing antibiotics and Fumidil-B. It was noted that the bees carried in quite a bit of pollen this fall but an estimation of the number of square inches was not made. A local beekeeper and I have discussed my wintering problem and he is fearful that I may have trouble with them because they are so heavy. He is afraid, since there is so much honey, especially in the second story, that they may not be able to avail themselves of it during our severe winter. He suggested that I remove a couple of combs of honey from the center of the second story and replace them with partially filled and drawn combs. I am somewhat fearful that by doing this I will be removing honey from a location where the bees may cluster in winter and, thereby, cause a lack of honey in severe weather when the cluster cannot expand to reach the surrounding abundant honey. What do you think?

A. We suggest that you reduce the lower entrance of the colony. This has the effect of keeping out the cold winter winds and at the same time will keep out field mice and rodents which otherwise would get into your colony. In addition, we recommend that in the top body, just under the hand hole in front, you bore a 1-inch auger hole. This will allow the bees flight during short warm periods in winter when their lower entrance might be clogged with ice and snow. Place a piece of ½ inch hardware cloth over the auger hole to keep field mice from entering.

We would not bother the top story at all. The bees will be eating up into the bottom portion of those fully drawn combs, and will establish their cluster upwards gradually during winter. However, since you are so far North, we would recommend that you at least wrap the colony with tar paper and/or provide a windbreak for the bees.

ENTRANCE REDUCERS FOR WINTER

Q. Is it important to use entrance reducers during the winter and when should they be put in?

A. Entrance reducers reduce the amount of cold air that can enter the hive and will keep the mice out if covered with ½ inch mesh hardware cloth. Mice will start to move with the first freeze of fall, coming in from the meadows, and many will make their way into a hive if nothing restricts them. They will go back into the corner of the hive, make a nest of grass, and do much damage to honey and the combs. An entrance cleat can be made by sawing a 1-inch board long enough to fit into the entrance, then make two cuts about 4 inches apart near the center of the board and ⅜ inches deep, then chisel out the part, making an entrance ⅜ x 4 inches. This will keep all mice out, except the shrew, which is somewhat smaller than the field mouse. Covering the entrance portion of the entrance cleat with hardware cloth will ensure that both are kept out of the hive.

VENTILATION HOLE

Q. Do you feel it advisable to have a small, second story landing board and a ¾ inch hole (the winter vent) in the back of the hive? This would enable the bees to work both sides of the hive easier, wouldn't it?

A. We use an upper entrance of one type or another, but we always place it in front of the hive beside the handhold. We close this entrance in the spring as bees only need one entrance through which to work. The upper entrance, although it may be used as a flight hole, seems to be the most valuable for upward ventilation.

TOP ENTRANCE

Q. What is the purpose of the top entrance during the winter?

A. A top entrance, such as a ⅝ inch auger hole just above the hand hold, will allow any moisture that is in the hive to escape, and also provides an outlet in case of deep snow or ice or dead bees blocking the lower entrance to the hive.

CLOSING THE TOP ENTRANCE

Q. I wintered my bees with a tarred paper wrap and a ⅞ inch top entrance. They all came through very well. What I would like to know is, now that they have started to bring in pollen, should I close the top entrances to keep the heat in the brood chambers?

A. You should close the top entrance in May but not to retain heat in the hive. Little or no heat is lost through this entrance. However, it should be closed to cut down on confusion when you are working the bees, and also to get them to use the bottom entrance. Less pollen seems to be stored in the supers if the bees use the bottom entrance.

WINTER SHELTER

Q. Would it be all right to place my bees on the south side of my garage for the winter? They would have to be set right against the white building so therefore would be warmed considerably on a sunny day. Would this variation in temperature decrease or increase their chances of wintering successfully? Would they require more than the usual amount of winter stores?

A. You could probably winter your bees successfully on the south side of your white garage. It is important that the bees get warm at intervals to move their cluster and even take a cleansing flight. It is also possible they would consume more food, but then again they may rear more brood in the early spring which should make them a better colony.

WINTER VENTILATION

Q. What is the best method of winter ventilation in our area (Wisconsin)?

A. The best system for you to use for ventilation during the winter months would be to reduce the main entrance, especially so that the mice cannot get into the hive. Then, an upper entrance can be made just above or below the handhold in the upper hive body. This should be a ¾ or 1 inch hole bored through and then covered with ½ inch hardware cloth to keep mice from getting in here too.

WINDBREAK FOR WINTER COLONIES

Q. I have 4 hives of bees from captured swarms last year. For wintering this year (Illinois), I had a dozen or so wooden storm windows. I built a three-sided windbreak with a glass top around each hive. It is open on the front so bees can fly on sunny days (facing south). This arrangement allows sun to hit hives from all angles, at the same time allowing air in front for ventilation. Is there a danger in bees becoming over active in brood rearing or any other chance of disease? I medicated for prevention of AFB and Nosema in the past fall. On recent examination, the bees were active and in all appearances healthy.

A. The windbreak you made for your bees was a very good idea. You would not want your bees to become overactive in the middle of winter; however, you would want to start your stimulative feeding around the middle of March for your area. One of the most important things in feeding in the spring is pollen or pollen substitutes.

PLASTIC COVERING KILLING BEES

Q. We never had any previous experience with bees. We obtained a colony of bees last summer. For the winter, my husband covered the entire bee hive with a heavy plastic sheet for winter protection. Now we found out all the bees are dead although they still had some honey left for food. My husband contends they died of starvation. I feel they died of suffocation. How important is ventilation? I am most unhappy over what has happened. Do you have any answers, please?

A. The use of the plastic sheet around the hive was very possibly the cause of the colony's death. Ventilation is especially important, and ability for the moisture in the hive to escape is important in wintering a colony. Moisture laden stale air in a wintering colony is extremely detrimental. Most beekeepers who winter colonies try to provide an upper entrance, even though very small, in the winter. This allows the moisture laden stale air to escape from the hive.

USE OF STORES

Q. Do bees use more stores in mild or cold weather?

A. Most likely the bees that are flying will consume more food than those which are quiet in the hive but that does not tell the entire story. If the bees are out in an unprotected area where the wind will strike the hive, especially the front of the hive, those bees will consume more honey than the ones which are in a hive that is protected from the wind. That is the reason why the bees should be located on a south slope and have a wind break to assist them in saving some of their winter stores.

BEES STARVING WITH HONEY STORES

Q. What causes bees to starve in the hive when they have honey stores around them?

A. Sometimes in extended periods of cold weather, the bees will have consumed all of the honey near them and due to the cold, cannot move in the hive to where they can reach the honey and therefore die. Usually a strong colony will be able to move upward to reach the honey stores.

WINTER STORES AND WINTERING ABILITY

Q. It has been my experience that an occasional colony will come through the winter with several times the amount of honey still left in the hive, as that of other colonies of the same strength. This also applies to the syrup left in the feeder cans of an occasional package of bees arriving in the north. Has this been a consideration in a breeding program?

A. Wintering ability and consumption of winter stores is considered in all of our breeding efforts for Dadant hybrids. In the first place, selections are made among the various hybrid combinations for those that not only produce the most honey, but also best prepare themselves for winter with stores available in the brood nest area of the colony. After wintering, evaluations are then made concerning the consumption of stores during the winter period, and even more importantly, during the spring buildup period when the majority of stores are actually consumed. Each hybrid released must be better than the previous one in relation to total honey production, preparation for winter, and the proper use of stores during the winter and the spring buildup.

DRONES IN NOVEMBER

Q. I saw several drones flying out of the colonies in November. I wonder if this is normal?

A. Usually when the bees are taking cleansing flights on a warm day and there is the discovery of some drones flying, it indicates that the queen in that colony is not up to standard and the bees have tolerated some drones in case she is disposed of. We would check that colony as soon as possible in the spring to see what the egg laying pattern of that queen looks like. Order a new queen and requeen the colony as soon as possible.

BEES LOST IN THE SNOW

Q. Can bees be stopped from flying out in the snow in the winter and being lost?

A. No. If you close them up, they will fight themselves to death trying to get out of the hive and if you put the colony in a building, you will find them all over a window or door trying to get out. They become confused and are lost. Bees need some means of leaving the hive on warm days for cleansing flights. Unless it is an unusual situation, there will not be an excessive number of bees lost.

DEAD BEES ON THE GROUND IN WINTER

Q. I am having some kind of trouble with my bees. I notice every morning around the front of my hives lots of dead bees lying on the ground. I also notice that when they fly out when there is snow on the ground, many can't make it back to the hive, and fall on the snow.

A. Yours is a common complaint. It is natural to find some dead bees on the ground in front of the hive after a period of confinement. Old bees die and are carried out as far as the bees can get them away from the hive. The ones you find on the snow may be those that got chilled and couldn't return.

There are arguments on both sides as to the value of these bees which couldn't make it back to the hive. Some say they are old and worthless; others that any live bee helps the colony. We're inclined to agree with the latter. We would like to see them stay alive and wish we knew a practical way to accomplish this. A friend covers the snow with cardboard to keep the bees out of the snow. Straw has been used by other beekeepers. These measures are practical only on a very small scale.

SNOW ON HIVES

Q. Will a snow drift over my bees cause any damage?

A. No. Actually, snow is a pretty good insulator against wind chill and the bees seem to be able to get enough air through the snow. If it is not packed solidly, they will be all right for the winter. We have heard many beekeepers say that they felt the snow around their hives saved the bees.

DEAD LARVAE IN FRONT OF COLONY

Q. For the past few days, we have had usually warm weather for this time of the year. This is the first time in about a week that we have had warm enough weather for bees to fly. I decided to take advantage of this situation and feed one of my colonies that I feared was low on stores. While brushing the dead bees away from the entrance, I noticed four dead larvae and yellow splotches all over the ground near the hives. The splotches appear to be dysentery but I cannot account for the larvae.

A. It is possible that the yellow splotches you observed could have been dysentery. However, they might have been just normal cleansing flight signs for the bees. This is normal after a period of close confinement. It would be even more pronounced if the bees had been confined for a month or longer.

It would be difficult from here to determine just what the cause of the larvae laying at the entrance would be. It could be that the larvae were detected as being diseased by the nurse bees and pulled out. However, this is unlikely as the bees would normally pull out the larvae in pieces. We had observed in the past that when a frame is placed outside of the hive for very long, the larvae, as they get hungry, will crawl out of the cells. If your hive was badly in need of stores, it could be that the larvae were suffering from the lack of food and that some of this "desperate crawling" has occurred.

EXAMINATION OF COLONIES

Q. There seems to be no life in my weakest hive since we had 25 degrees below. When can I open it up and see? If the bees are dead, can I leave the combs as they are until I get a 3-lb. package April 1st? Could I use the honey in the combs if there is any left? Do I need to seal the hive to prevent wax moth damage?

A. It would be desirable for you to wait until the temperature gets to at least 40 degrees above zero before opening the colony for inspection. The warmer, the better, of course. By opening the colony in too cold weather, the heat within the colony escapes, and you endanger the survival of the colony.

If in fact the bees are dead, the combs should be cleaned of dead bees to have them ready for the package bees. The honey that is left could be used by the package bees.

There may not be so much danger of wax moth in the hives now as the wax moth doesn't really operate until the temperature warms up considerably. The invasion by mice would be of more danger since they would quickly destroy the combs. Another problem may be robber bees trying to rob out what honey is left.

WORKERS

LAYING WORKERS

Q. Do you have any specific information on how to rid a hive of laying workers? The bees were shaken off the frames some distance from the hive but the laying workers must have made it back to the hive.

A. There are a number of possibilities that can be followed with relation to getting rid of laying workers and requeening the colony. You do not say how far away you took the frames with the bees on them when you shook the bees off. Normally, the farther away you go, the more likely you will be to lose the laying workers. Quite often these are somewhat heavily developed in the abdominal area and, therefore, find it more difficult to fly back to the colony.

If there are other colonies of bees, there are several things you can do using them. One would be to move the colony which has the laying workers, after having shaken the bees off the combs, to a new spot in the yard. Then, a number of combs of emerging brood could be taken from the other colonies and placed in the center of the hive. These combs should have the bees on them, but you must be careful not to move the queen from the colonies they come from. The push-in cage is the most foolproof method of requeening a colony under adverse conditions. We would suggest using this method and after shaking the bees off the comb, place the new queen on the comb and put the push-in cage over her within an area in which the young bees have partially emerged and are continuing to do so.

If the general age of the bees in the queenless or laying worker colony is getting too old for requeening, then it might be a good idea to unite this colony with a queen-right colony nearby. This can be done by using a sheet of newspaper over the top of the combs of the queen-right colony. Then, make small slits in the sheet of newspaper and place the hive body from the queenless unit on top. The bees will soon eat their way through and mingle together and make one strong colony of the two units. The bees from the now strong colony will dispose of the laying workers.

LAYING WORKERS

Q. I have kept bees for six or seven years as a hobby. Here is my problem. I hived a small swarm of bees that had settled low and it was easy to get them into the hive. I put them in an 8-frame brood chamber with six frames of foundation and two frames of young bees and honey which I had taken from the parent hive. Later I looked in the hive and couldn't find a queen. The foundation was partly drawn and had honey and eggs, and bees were bringing in pollen. About half the cells had eggs in them, two or more eggs in each cell. Also on three frames they had built queen cells up on the frames. These cups all had eggs in them, more than one in each cell and as many as twelve eggs in one cell cup. Now, do I have a laying worker? Could I reunite these bees back with the original colony they swarmed from? What would be the best thing to do with them?

A. A queen will often lay more than one egg in a cell when she has only a few cells available and a young queen will also do this but it sounds as though you have a laying worker. Laying workers are a problem and requeening is difficult. Unless you have a fairly strong colony of young bees, it would be difficult to maintain them as a separate colony and we would suggest that you unite them with a queenright colony, using the newspaper plan.

LIFE OF A WORKER

Q. If the life of a worker bee is so short, how can they live all winter?

A. The life of a worker bee is usually shorter in the summer, depending on how much work they are doing — brood rearing, flying, or heavy foraging which soon wears out the wings. A bee that is in a cluster can live much longer than the summer bee and the young bees going into the winter cluster can survive until the young bees start coming on the following spring.

LIFE OF A WORKER

Q. How long does a worker live?

A. The length of life of the individual worker honey bee varies tremendously at different times of the year. Experiments with marked bees have shown that in a normal queenright colony in March the average expectation of life on emergence is about 35 days, but in June it has become reduced to an average of about 28 days, about 9 of which are spent foraging. On the other hand many of the workers that are reared during September and October live throughout the winter. There is a record of such individuals living for 304 days. The three factors would seem to be: (1) the amount of pollen consumption; (2) the amount of brood-rearing; and (3) the amount of foraging outside the colony.

INDEX

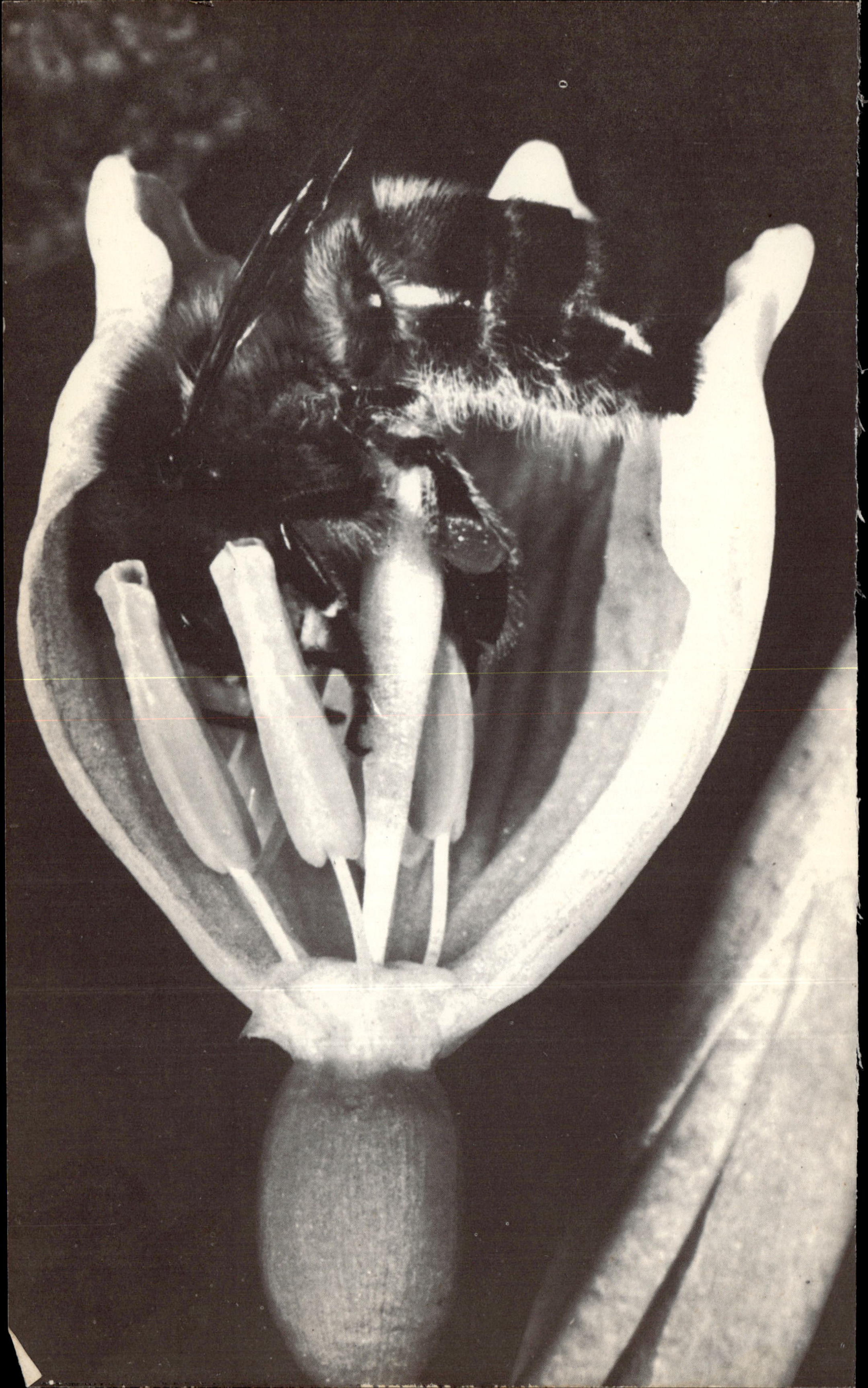